The Hope, Love and Resilience Expedition

By Mark Ivancic

Copyright © 2020 Mark Ivancic
All rights reserved.
Published by Mixx and Koki Publishing
ISBN: 978-1-989849-07-1

DEDICATION

This book is dedicated to my daughter Megan, my biggest supporter in my life; who taught me that I'm a student for life.

FOREWORD

My Friend Mr. I

I have known my great friend Mark Ivancic for close to 25 years now. We first met while playing hockey – something that we both are passionate about. As we got to know each other a little bit more - we realized we had many things in common. Both of us played university hockey and graduated with a Bachelor of Education degree.

From that first skate together and throughout the years we remained close friends. Not only because of our love for hockey, but our philosophy towards coaching and providing positive environments for today's youth. As dedicated dads/coaches, we would often discuss the challenges and rewards we experienced on the hardcourt or ice surface. Mark is a great competitor and loves challenges. He turned a small school girls' basketball program into a perennial powerhouse program known throughout the province.

From our talks – I knew Mark was an innovative person, looking to take teaching in a different direction than the - same old blueprint. With the support of his principal, he started a program at his school called Life Skills. Within this program – Mark would be able teach outside the box and provide key skills to help students navigate the game of life. No one would be better suited to this task.

It was clear thru our discussions, that Mark was creating an amazing learning environment for his students. I was lucky enough to view this firsthand when I was invited to speak at an assembly at a Career Day presentation he was conducting.

What an eye opener! I was blown away by the atmosphere in his classroom. I believe three classes of students were assembled and they were the best behaved group of elementary students I have ever been around. Mark did the introductions then picked up his guitar and started playing a song. Before long – every student was singing with him. It was amazing to witness, there was such a connection with "Mr. I" and this group of grades four-five students. It was then that I realized my friend was no ordinary teacher. This just doesn't happen in our school system. It was about this time I started thinking about how lucky these students are to have such a positive - invested teacher.

Throughout the years, whether we were at a restaurant, concert or sporting event, on cue, former students would recognize him and make their way over to talk. They would always smile and relive the time they had spent in one of Mark's classes saying it was the best experience of their life! I thought to myself, this must be what sports heroes face when they go out for an evening.

To Mark's credit – he remembered all their names! The common theme here was how much they enjoyed him as a teacher and the life lessons they learned in his class.

If you are a buddy of Mark's, you soon realize that he has a very high humanitarian compass. No one I know has worked so hard for causes. He has unselfishly donated time, energy, and passion to help others supporting great causes such as 'The Children of Equator', 'Band against Bullying', and 'Power in Me'. He has helped and touched so many people in such a positive way in a way that only Mr. I can!

Life Changes

To this day, I still can't believe it happened. Mark was freshly enjoying retirement, looking forward to the next chapter of his life. He was in amazing shape, working out daily in the gym and playing hockey a couple of times a week. Two years previous, he had fiercely rehabilitated from a seriously blown knee injury suffered when he was playing floor hockey with students at lunch time. No offence to Connor McDavid, but this injury was way worse - tearing all three ligaments - MCL, ACL, and PCL.

A life altering event occurred in April 2018. As we prepared for a tournament in Banff, I played on a line with Mark and Brad – the two fittest guys in our age category. Mark was flying that night and I was having trouble keeping up with the play. I remember thinking to myself – What great shape Mark is in! After the game we headed next door to Boston

Pizza to grab a bite, beverage and some round table discussion. After we ate, Mark got up and said he felt a migraine coming on and was going to call it a night. If we had only recognized this as one of the signs. Mark suffered a debilitating stroke and was operated on to remove a clot. He has paralysis on his left side. How could this have happened? He was one of the fittest guys I know. He climbs mountains in the summers and works out in the gym daily. We were all in shock!

Mark was in the battle of his life but in typical Ivy [Mark's nickname] fashion he refused to let the stroke take control. He made two key plans while in rehab at the Glenrose Hospital. First to write a book about his situation and the challenges that face people who have suffered strokes. Second, to plan a hike up Sulphur Ridge with former students and friends. It is one thing to write a book when you are 100% healthy, but try writing it in verse form only a month after you suffered a stroke! Pretty amazing to say the least!

My friend was bedridden with paralysis on his left side when he mentioned his idea to climb Sulphur Ridge. This hike by the way, is a 10 km trip with a very good vertical climb. This is a special place for Mark because in his Life Skills class he had taken so many students to the top. As much as I want to believe he could do this trip, I did have my doubts.

Fast forward four months and 28 of us are at the bottom starting point of Sulphur Ridge in Jasper. Not everyone made it to the top, but Mark Did! What an amazing feat for him and everyone else on that hike! To this day, I still can't believe Mark climbed to the top of the ridge just over four months after surgery for a clot in his brain! He has no idea how much of an inspiration he is to me and so many others!

The Idea

How did the idea for another book come about? Last spring, I picked Mark up (we are in each other's bubble) and we headed out to the farm. On the ride there, we talked about the uncertainty the world is facing with the pandemic crisis. On the news every day – nothing good only how bad things are -

and how many people are passing away from Covid. People are isolated and feeling very down. Not since the early 1900's has the world experienced anything like this. People need hope and positive messaging. "I think you have a great message to pass on to folks" I said to Mark. "I see how much you have positively influenced former students – why couldn't you do it to a wider audience?" People need good things to read now more than ever!

The seed was planted! The writing started and the ideas flowed. I have read the drafts and Mark doesn't know this, but the words have really helped me personally. Like many during this pandemic, the mind isn't quite as strong and there are times when I have felt down and allowed myself to question decisions I have made in the past. Reading the words from his manuscript helped me to overcome the doubts and replace them with good positive thoughts. I strongly believe the words, ideas, quotes and heart-felt comments in this book will serve to help many. It is a fun read and one cannot feel anything but good afterwards.

To quote a few words from a song written by one of Mark's favourite artists:

HARD DAYS ~

If you never had hard days
If you never had a heart break
Never had more than you can take
Or carried the weight on your shoulders
Would you feel like you earned it
Would you live with a purpose
Or ever know your own strength
If you never had hard days

by Brantley Gilbert

By Dave Souch [Souchie]

Well, what a kind and wonderful foreword from my bud, Souchie [Souchie is my nickname for my great friend Dave Souch and you'll hear more from him later in this book]. To be honest, I think what impressed him the most with my interactions with former students was when we would be out enjoying refreshments and we would get a free round sent our way even if it didn't happen as often as he would like to believe. He remained adamant I complete this project often acting as my pillar of support when I wavered never allowing me to entertain the thought of not finishing for which I am extremely grateful and hopefully, when all is said and done, you feel the same way.

THE INTRODUCTION

This is the story behind the story and the genesis of how this book came to be. It all started over a quarter century ago and I am not sure just exactly when. I was personally going through a rough patch in life and started reading all kinds of motivational and self help books. From there I started dabbling in becoming a motivational speaker doing some small engagements here and there, but really gaining a lot of satisfaction and enjoyment in doing this.

One day, I was chatting with a fellow teacher who suggested I take what I was doing on the side and offer it as an options class. I suspect the real reason why he planted that seed was that we were providing limited choices for options. There was a core group of grade nine students who coloured outside the lines who some teachers may have thought to be a formidable and intimidating group to rein in.

Right from the start I was thrilled with this opportunity and this group proved to be the perfect match for me, inspiring me as much as I did them. Not knowing what to call the class, I decided on calling it Life Skills, thinking if we were emotionally strong and possessed positive belief systems, we would have the skill set needed to move forward to achieve personal success in life and to overcome the obstacles it inevitably throws our way.

I had no program to follow so I just tried different strategies and activities as I muddled along, but they were a wonderful group who bought into what I was selling right from the start.

The one thing I did that seemed to go over well was the use of motivational quotes and the stories I used to enhance the message. As you shall see, many were personal because that is one's point of reference, but I wanted to come across as genuinely sharing my failures and my successes.

Based on their positive reactions, this would serve as the winning formula I would employ for years to come. Even teachers started to notice a positive change in the students' attitudes and behaviours. We ended the year in spectacular fashion as a boy in the class was diagnosed with colitis and the rest of the class were determined we need to do something about it. We did a wake-a-thon fundraiser at the school, raising over $2,500.

When June rolls around, the grade eight students have to select their option choices for the following school year and my program is the hot ticket item. I suspect our fundraising event had a lot to do with that. My principal at the time, Mr. Roger Murray, an amazing man who I hold in such high esteem, who supported me and believed in me regardless of how far I wanted to push the boundaries; makes an executive decision and mandates this is going to be a permanent fixture in our school and that all grade nine students will be required to take this class.

He had so much faith in me, for which I will be eternally indebted, which goes to prove the impact we can have on others when we believe in them and have their back. So much so that when one year we had a problematic group of grade sevens, his solution to the problem was that Life Skills will now be offered to all grade eight students as well as the grade nines.

Challenge accepted because I was thrilled I would have the opportunity to be in the position to do something I was passionate about and have it as part of my job and my work day and I benefitted the most because the more I talked it, the more I had to walk it, which I would like to believe made me a better person and teacher. To make a living doing something you thoroughly enjoy is better than awesome!

After having my stroke, I suffered from left side paralysis which meant I had to forsake on all my passions and seemingly future ambitions. And here is the first quote from Albert Einstein who stated:

> *"The definition of insanity is doing the same thing and expecting different results!"* ~ Albert Einstein

So it's time to take action, which is why I am writing this book. Your first step is to grind your way through it and actually try to implement some of the suggestions and strategies. I can't remember where I got this next quote from, but it reinforces the need to making things happen in our lives.

> *"The best case scenario is making the right choice, the second best scenario is making the wrong choice and the worst case scenario is doing nothing!"* ~ Author Unknown

Read on and accept the 1% challenge and see where it takes you. This book is a series of quotes, stories and lessons the students learned from me and most importantly, the lessons I learned from them, from the Life Skills Class I taught for many years.

I hope you enjoy reading this book as much as I enjoyed writing it.

THE JUNIOR HIGH LIFE SKILLS CLASS WE ALL NEEDED TO TAKE

As outlined in the introduction, this book is totally based on the Life Skills class I offered to junior high students and I have attempted to replicate it onto paper as close as possible. So as you read this, remember a lot of the material is presented as if you were a young teenager filled with unlimited potential and endless possibilities which is a mindset we can lose but should try to maintain. So as you read, try to embrace the material as if it were the first time you were exposed to it rather than treating it as redundant and a collection of cheesy bumper stickers and Hallmark cards.

My initial idea for this book was to call it THE 365 DEGREE CHALLENGE thinking I would have 365 entries. When I was recovering from my stroke, it seemed like if I started my day off by reading something positive; it put me in a better head space for the remainder of the day.

The thinking here was to have an entry for each day of the year since, you the reader got your hands on this book. The challenge being whether or not you could grind your way through 365 pages of my rantings. Alas, fear not, this proved to be too big of a challenge for me and I bailed, but I still do believe starting your day off with a page or two of this book would still be a pretty good launching point and most beneficial.

THE ONE PERCENT CHALLENGE

My thinking here was to get the reader in the mindset of how to improve the quality of each day by the slightest of margins, hoping this book could prove to be a catalyst in making this happen. With continual small increments, hopefully impacting someone on a much larger scale in the long run. You will notice throughout the book, I still make references to this title because it is something I strongly believe in and it can make a significant difference in someone's life. I know, because I lost my way and my mojo, but in re-immersing myself with this material has brought me back to a place I needed to be at in order to move forward and enjoy life like it is meant to be for all of us.

STROKE RECOVERY AND WORKING MY WAY BACK TO PAR

Nope. I often reference my own personal predicament, but it would seem to me many people are recovering from a variety of difficult situations so why keep the scope narrow? This concept led me to my next possibility, LEARNING HOW TO HEAL. Through my personal experience, healing involves more than rest, medicine, therapy and physical treatments. You may try to approach it with logic, but there is an emotional component that needs to be addressed after going through some type of trauma. Until you do this, the rest is moot. So in essence, we need to learn how to heal emotionally before we can truly move forward.

THE LITTLE BOOK OF GOODNESS

As I started to receive powerful and personal submissions from friends, goodness seemed to be the common theme and it just resonated so strongly.

And finally, THE HOPE, LOVE AND RESILIENCE EXPEDITION. I could have used the word journey but to me expedition signifies a journey with other people being involved plus it implies a mountain adventure which is something I am rather passionate about and a focal point I keep coming back to. As well, this book is rife with examples of hope and love and who can have too much of that?

IF YOU ARE LOOKING FOR SELF IMPROVEMENT, JOIN THE BAMMA RAMMA SLAMMA JAMMA MOVEMENT!

I am going to start by giving people the insight into what the bamma ramma slamma jamma represents. It was my self-invented catch phrase and in my post stroke book that I wrote.

I would say it when I was excited and the students would be delighted. And that's all it took to get it ignited!"

I guess that would be a good place to begin. To me the expression represents being in a happy place with others and experiencing a moment that brings everyone pure joy. Like when you create a positive memory or accomplish something when you had to push yourself and may not have thought possible or simply sharing a smile with family and/or friends. To me the bamma ramma slamma jamma represents a celebration of life. Moments that will live forever and serve as the driving force for us to create new memories.

I could go on and on, but we all have our own perspective to what constitutes a bamma ramma slamma jamma moment. The key here, is how can we make it happen? It involves stepping up and taking action when we are consumed by self-doubt, taking a calculated risk, spending time with others, trying to find positives when all is seemingly lost. Only you and you alone knows what brings joy into your life. So make it be and fill your life with the joy of the bamma ramma slamma jamma!

MONSIEUR JE NE PEUX PAS IS NO LONGER WITH US

I was going to be evaluated by my vice principal Dave when I noticed my grade four French Immersion students were having difficulty in properly arranging the ne and the pas in their sentences. I was reading 'Chicken Soup for the Soul' at the time and there was a story about a teacher who had her students write on a sheet of paper all the things they thought they couldn't do. She gathered all their papers and they proceeded to have a funeral service for 'Mister I Cannot' which is the English translation of Monsieur Je ne peux pas. At the end of the service she explained how 'Mr. I Cannot' was no longer with us as they dug a hole and buried all the papers. For the rest of the year they were not to say 'I can't do something' because he had passed on and no longer existed. I thought this was brilliant and could serve as the perfect lesson to which I was to be evaluated on.

I explained how to properly position the ne and the pas when putting a sentence into the negative. Following what the teacher in the book had done, I had my students write in French all the things they thought they could not do.

I got a shovel for the burial, but it was November in Alberta and the ground was frozen solid. I had to adapt and told the class instead of a burial we were going to have a cremation. Having over twenty sheets of paper in my hands, I had a student go get the trash can in the class so I could drop the burning papers in there without scorching my fingers.

Of course they didn't empty the can first and they brought out the one that was made of rubber. As I was finishing up the funeral service, I lit the papers and they burst into flames. I had to toss them into the trash can which ignited whatever was in there. The fire was contained, but the rubber sides started to melt and ten-year-old boys being ten-year-old boys started poking the rubber, disfiguring the shape of the bin. Needless to say, when it cooled off, we had the most unique looking trash bin in the history of education and the words je ne peux pas were never used for the rest of the year because we had a visual to remind us. I am not quite sure what type of evaluation I received, but Dave and I still chuckle about my lesson to this day.

The bottom line is, we all have this alter ego who tries to whisper self-doubt to us or who slows us down. Imagine Superman without Clark Kent, Batman without Bruce Wayne, Spiderman without Peter Parker and you without Monsieur Je ne peux pas and the amazing super feats you were destined to accomplish because there was nothing holding you back?

> *"Whether you think you can or whether you think you cannot. You are right."* ~ Henry Ford

I think the previous page served as the perfect segway for this quote. With this quote, I liked sharing a story passed on to me by my former colleague and climbing partner Chris who was a popular figure in the building. He had gone rock climbing on a

5.9 route and when he got to the crux, the most difficult part of the ascent; he kept popping off, eventually having to retreat to try again another day.

Several weeks later, he returned to the scene of the crime. He misread the guidebook and thought he was on a 5.7 route. When he arrived at the crux, he had it in his head it was a 5.7 route and believed in his heart that there wasn't a 5.7 route he couldn't master, so he easily navigated his way through the difficulty and it wasn't until he got back down from the climb when he realized the mistake he had made and which route he had just successfully completed. From that point on, that crux would never prove to be an obstacle for him. It just goes to prove, when we put ourselves in the right frame of mind, success is within our reach.

One has to look at the role sports psychologists play in today's athletic scene. As a former university hockey player, I can recall times when I was on a roll. Everything around me seemed to be moving in slow motion and I could attempt things beyond my normal range because I believed I was that good. And then you hit a slump. The puck was like a ticking time bomb and I couldn't get it off my stick soon enough. And here is the point. I possessed the same physical skills so the only difference was how I viewed myself. Often in life, we are so close to success only to allow self-doubt to enter in our heads and convince ourselves otherwise.

The key to success is often positive self talk. There were times in my scrambling career where I had to ascend up a daunting rock face without protection; note to self, do not say that in front of junior high students. BTW – Scrambling means, getting your ass up a mountain any way you can and literally scrambling on your hands and knees. I would convince myself there were people much less athletic than me who had successfully completed it and surely I could do so as well. To be honest, this only served to put me in more precarious situations than I needed to be in, but I sure did enjoy myself.

Anthony Robbins wrote a book called 'Unlimited Power' which I used as a primary resource because it was filled with juicy morsels. His diligence and insights are invaluable and I would share this passage every year.

> *"The birth of excellence begins with our awareness that our beliefs are a choice. We usually don't think of it that way, but belief can be a conscious choice. You can choose beliefs that limit you, or you can choose beliefs that support you. The trick is to choose the beliefs that are conducive to success and the results you want and to discard the ones that hold you back."* ~ Anthony Robbins

It is amazing how often 'what' we feed our brain comes to fruition. To prove my point, I would ask the class, "How many of you have pretended to be sick in an attempt to not go to school that day?"

I asked them to raise their hands. Of course, the response was consistently unanimous. I asked them to keep it raised, if after throughout the day, they actually didn't feel that well? I could sense by their reactions they realized I was onto something and may actually know a little thing or two.

Doctors have been administering magical sugar tablets for decades to children explaining how it would cure whatever was ailing them. I would offer example after example of people being led to believe something and how it would become their reality.

There is a reason why lie detector tests are not permissible in a court of law. If you continually feed your brain the same message, you literally go into a state of making it your perceived reality and can beat the test even if you are guilty. I feel comfortable sharing this tidbit of information in how to beat the system because I do not think people with nefarious intentions would be taking the time to read this book.

Right now, I am in the process of convincing myself I will completely heal from the damage I suffered from my stroke because this is holding me back from living the quality of life I want. Even though I had people who have expertise in this area tell me otherwise; I tell those around me, "I WILL HEAL!"

Just by saying those words puts me in a better head space and provides me with hope. I realize there is no guarantee I will fully recover, but it feels a lot better focusing on the positive possibility rather than dwelling on what I cannot do. There are medical miracles all around us, so why not me? You have to ask yourself, what are the faulty belief systems you have that are preventing you from enjoying the quality of life you deserve?

I have received a wide range of reactions over the years from students confessing of engaging in suicidal thoughts, only to be convinced otherwise to those who disliked it for whatever reason. Maybe it was because I came across as too preachy, talking like I had all the answers, but I know I am not the first or only teacher who conducted themselves in such fashion.

> *"The power of love will always be greater than the love of power." ~ Unknown Author*

Have you ever noticed when there is a catastrophe, whether it be personal, regional or even global, there are those who rise to the occasion by going above and beyond adding a ray of light to a dark and gloomy situation who demonstrate selflessly the best humanity has to offer?

Often, they receive no fanfare or recognition for stepping up, but they provide us all with the glowing example of the potential and ability we possess and what we should all aspire to be like because they can make the most difficult of situations bearable, compared to the toilet tissue hoarders.

I would always say your true colours come out when things become difficult because it is easy to act cool and behave like you have your act together when life is running smoothly, but how you respond when things turn sour reveals what you are truly made of.

The real reason this quote resonates with me is because it serves as another catalyst for the creation of this book. My wife and I went to a local pizza establishment and our server is a former student. She brings up the class and out of the blue, she states her favourite quote.

I am blown away by this exchange and feel so remiss why I didn't ask why this particular passage stood out for her.

I didn't consider this to be one of the more meaningful passages, but it got me to thinking what other quotes or lessons stood out for different students.

I put out a call to former students to share with me the most impactful one that stood out for them. As you shall see, I have included them sporadically in the book for more credibility with you, the reader. Their stories are humbling and heart warming and hopefully there will be something in this book which will strike a chord with you and help propel you forward.

SATCHEL PAIGE

I would often share stories of remarkable people and not only was Satchel Paige, a remarkable man, but he is responsible for two of my all time favourite quotes. He is one of the greatest pitchers to play baseball who no one knows about. His stats are truly incredible as he was basically unhittable in his prime.

Unfortunately, due to the colour barrier in major league baseball at that time, he could only pitch in the Negro Leagues which was still an exceptional calibre of baseball. He eventually got the opportunity to play in the "big leagues" when he was well past his prime.

When asked if he was bitter about how this played out, he replied, *"IT'S MIND OVER MATTER, IF YOU DON'T MIND, IT DOESN'T MATTER."*

The lesson being, there is no point dwelling in anger when things are not in our control. Do the best you can and keep moving forward.

People figured he was probably well into his forties when he finally got the opportunity to fulfil his dreams. When asked about it and not wanting age to serve as a deterrent for this chance, his response was, *"IF YOU DIDN'T KNOW HOW OLD YOU WERE, HOW OLD WOULD YOU BE?"*

I embraced this adage as it became my personal mantra. I just love to play so I never wanted a silly thing such as age to prevent me from having fun. And the key to the fountain of youth? Physical fitness. The more I trained, the more I accomplished and the better I felt about myself.

> *"No panic, no fear, no disaster!"* ~ Unknown Author

This is the quote I am probably most associated with by former students. I believe I read it in John Krakauer's book, 'Into Thin Air'. It describes the deadly infamous mountain expedition on Mount Everest.

A group of climbers were huddled together with zero visibility and in the middle of a hellacious storm with no idea where their tents were which could provide refuge and the chance for survival. My apologies if I botch what exactly took place, but this is my version of what happened and is the story I shared over many years.

For a brief moment, there was a break in the clouds and one climber was able to identify the north star and with that he was able to orient to where he believed the tents were situated. He instructed the group to follow him and that he could lead them to safety.

When a female climber was asked why would she risk her life and place so much faith in him. She explained how in the midst of all the chaos and in a life or death situation, he remained so poised and calm. He led them to the tents, saving the lives of several climbers. When asked how he was able to maintain his composure under such duress, he explained his motto of, *"no panic no fear, no disaster."*

Words to live by as we find ourselves in the grips of a global pandemic! I would cite this quote often having to put it to the test on my own mountain adventures. I probably really stressed the value of such a wise adage on the last night of our grade six camp trip.

The setting was very rustic, to say the least, as we were still using outhouses. On the last night I would tell a fabricated story of Arnold Poopchuck who was apparently still at large. Not wanting to have to air out any sleeping bags on the bus ride home, I kept reiterating the importance of keeping cool when your mind may be racing towards worst case scenarios.

I even had a former student reach out to me explaining how she was scared on her wedding day not knowing if she could bring herself to walking down the aisle. After a deep breath and a few repetitions of the phrase she managed to pull through. All I can say is that after thirty plus years of marriage maybe a little bit of panic and fear isn't so bad after all.

> *"If you change your attitude, you will alter your altitude."* ~ Unknown Author

Sometimes in life we need to restructure our take on a situation in order to make things more positive or even bearable. I recently applied this tactic in my life and it has made a world of difference. After suffering my stroke and not being able to move my left arm or having the leg strength and balance I had before; I was preoccupied with all the things I could no longer do.

Someone had commented on Facebook how I was an amazing man, but I felt so far from being amazing based on how I was handling my current situation. Then it dawned on me how I was now in a position to be truly amazing.

What if I turned this around and treated it as an opportunity to over achieve, to do things most people and even experts did not think were possible? Four months to the day after having suffered from a stroke, I got to the top of a mountain ridge with family, friends and former students. It is all described in my first book, AT THE STROKE OF MIDNIGHT AND I'M STILL HERE BAMMA RAMMA SLAMMA JAMMA.

I realized I could achieve so much more if I pushed myself, but I knew this was not going to be easy. I grabbed my Scramble book to see which new mountains I could ascend that I had never been on top of before. I started going back to the gym and began the process of rebuilding the strength in my legs.

This enabled me to interact with people again, further adding fuel to my new internal fire. It was like I now enjoyed a new found purpose in life and I knew once I scaled those peaks, I would be ready to tackle new challenges. These challenges excited me, thinking of the potential adventures that were awaiting me. In my eyes, the change this made in my life was unbelievably remarkable!

Of course, the gym was shut down due to the pandemic, but regardless, I still managed to hobble up to the tops of four summits this summer. When I scaled to the top of Mount St. Piran for the first time in my life, it was the closest I felt to being my old self in a long time. This has motivated me to take on more new exciting challenges. I have always believed life should be an adventure expedition and my new circumstances are proving that to be true. I had used this strategy before and it proved to be most effective.

When I was a student and got a job at a local mine, I found the work to be tedious and mindless; like maybe I was wasting eight hours on my lifeline every day. Needless to say, I was miserable and doing my share of complaining. When I restructured my thinking to treating this as an opportunity to get paid for working out, I began to volunteer for the more gruelling and strenuous jobs.

Not only were my work days easier to take and I was getting more fit, I also became much more productive. To the point that near the end of the summer, they started laying off students because there was not enough work to do. The full time guys stepped up for me saying they still required me to help them. I stayed on for the rest of the summer and a valuable lesson was learned.

On a side note. In retrospect, I never truly showed my gratitude for the one guy who really stepped up and vouched for me to stay on. Unfortunately, he has passed on and apparently by his own doing. I cannot help but wonder, if I had said something would that have made a difference? I am pretty sure my words would not have played that significant of a role in the final outcome. But I would always tell students, especially when it came to bullying, you are either part of the problem or part of the solution.

We have opportunities in each and every day to be part of the solution through a kind word or a thoughtful gesture. Now the question is, if more people had stepped up and acted like part of the solution, would this person have felt more valued and the outcome be different?

We can all snap a single pencil in two, okay I can't right now, but put a dozen pencils in your hand and see how far you get. The point being, if we all step up and make compassion and kindness the norm, how much stronger are we making others without even realizing it and just by regular small gestures?

I have no illusions this book is the answer for all of the woes people are struggling with because that would be too much of a burden, but I do take solace in hoping that it will be part of the solution in some small way. Point being, we can all find a small role to play in being part of the collective solution.

THE GOLDEN RULE- "HOW YOU TREAT OTHERS IS HOW YOU WILL BE TREATED!"

Every major religion has their own take on the golden rule which usually reads like, *"Do unto others as you would have done to you."* True enough, but everyday we are modelling how we want to be treated. Without fail, the grade fours who complained the other children were being mean to them, were the ones who were most often those who were more aggressive verbally and physically.

From my humble observations, it seems to remain that way right into adulthood. In presentations, I would ask students to raise their hand if they had told someone to shut up whether jokingly or out of anger?

Inevitably almost every student would have a hand in the air. Then I would ask them to raise a hand if they enjoy being told to shut up. Of course, no hands were raised. And if someone was being smart and raised their hand, I felt compelled to say, if you enjoy it, please shut up for the rest of my presentation and we will get along just fine. What? I was just trying to bring them some enjoyment because they liked it. They usually got the point how they didn't enjoy being spoken to in that manner, but were guilty of having done so.

It's quite an easy concept, if you want to be respected, you have to be respectful. If you want to be liked, you have to be likeable and if you want to be loved, you have to be loveable. I was not ready to accept the standard junior high excuse when they did something wrong; that it was just the way they were and the world should simply accept them for their faults and let bygones be bygones because of it.

SOUCHIE'S SIGNS

Souchie [my friend Dave Souche who wrote my foreword] has an acreage with over seventy acres of land out in the country. I would regularly go out with my dogs, that was until two trips to the vets to remove porcupine quills! His place is so serene and scenic, enabling me to get my nature fix, which we all need. The whole environment is so rustic and soothing to the soul. In his main living area, there is a painting on old burlap with just words in a wooden glass frame.

The words are the title of a Tim Magraw country song, "*Always Stay Humble and Kind.*" It seems like no matter where you are sitting, your eyes are constantly drawn to the sign sending a subliminal message to the brain. And it must work because Souchie is as humble and kind as there is. Fill your walls with images and sayings that bring joy to your heart.

Isn't it interesting how easy it is for us to stereotype people? Somehow if a group is classified as different, we may feel a need to devalue their beliefs and what they stand for, giving the group you associate with a false sense of superiority. I think the pandemic has demonstrated how we are all in this together.

The reason for this rant is that by now many of you think you know what Souchie is all about because he lives in the country, wears cowboy boots and listens to country music, he must be a REDNECK! Which brings me to his other sign. As soon as you enter his property, these are the first words you will read, "DUE TO THE RISING COST OF AMMUNITION, THERE WILL BE NO WARNING SHOT FIRED." YUP, He is a redneck! And very proud of it because he believes it entails being humble and kind with a few loose screws if need be.

TAHLON'S STORY

I put it on Facebook for former students to reach out to me with their favourite quote and why it resonated with them. The responses have been overwhelmingly kind and positive and I thought it would be a good idea to share a few starting with Tahlon.

Growing up, he spent a lot of time on his own with limited adult supervision which usually does not bode well for a child. But Tahlon was highly motivated and passionate about sports. He bought what I was selling early on and it was my pleasure to be his basketball coach in grade nine. He was successful in a variety of sports all throughout high school and enrolled in a golf management program in college.

I knew he was an assistant pro at a golf course in Edmonton. He came by the school one day, much to my surprise to drop off a thank you letter before travelling to Australia where he was going to work as an assistant pro for the next six months.

As thrilled as I was to receive the note, it was even more exciting to see this tremendous young man fulfilling his dream, giving me a positive story I would share every year with classes. He is now the head pro at the Fairmont Jasper Lodge Golf Course. Nestled in the middle of the Canadian Rockies and arguably one of the most scenic golf courses on the planet!

It pleases me to no end to be able to share his story and here is what he sent to me.

I don't even know where to go with this because there were so many of those moments in that class that gave me so many outlooks on life. The one that truly sticks out the most to me would be the 'lifeline' example. Comparing the ups and downs you go through in life versus a single flat line with no peaks or low points. You asked the question to the class, "If you were on a hospital bed and you were looking at your cardiac monitor, which graph would you want to see, a flat line or a moving lifeline with high points and low points?"

This taught me that no matter what the situation is, to stay strong as no matter how bad things can get you can always climb your way back. ~ Tahlon

> "Excellence is not a singular act; it's a habit. You are what you repeatedly do." ~ Shaq O'Neal quoting Aristotle

The 1% Challenge is all about focusing on what we do on a daily basis. As discussed on the Daily Delight entry. It is all about training our brains to consistently respond in a positive manner so that it becomes an automatic reaction without even having to consciously think about it, as it just becomes natural defining who we are and how people perceive us.

The first step is understanding you are excellence. It is in each and every one of us and it is up to us to bring it out and show it to the world. You need to believe you are excellence because you are the best you on the planet. Unless there is another you in a parallel universe. And you need to believe each day is filled with blessings and excellence because it is all around us and the practice of daily delights will reveal themselves to us.

Cautionary note: expecting excellence from ourselves can serve to be a dual edged sword. I have been described by those close to me as my own worst critic.

When asked about how I felt I did after a performance or a presentation, my standard response is, "I could have done better." I have high expectations of myself and it's not that I was being overly critical, but by keeping the bar high, I kept improving. I would reflect on the excellent moments I might have enjoyed and try to figure out how I could make it more consistent throughout the entire performance.

Each day is like that. We have excellent moments and each new day gives us the opportunity to start with a clean slate with new chances and opportunities to strive towards excellence. I am sure you would never find an interview with Michael Jordan or Wayne Gretzky being asked about their performance after a game with them replying, "It was great even though I was just average today."

Like anything in life, striving towards having daily excellence is possible with constant repetition and practice. Being ordinary and just fitting in the pack never appealed to me. Whenever we received new junior high students, I always enjoyed the looks on their faces after their initial Life Skills classes because I could tell they were thinking, "You aren't like the other ones are you?" Not that I am suggesting that I was excellence in teaching, but that it is okay to be different because I definitely was that.

> *"It's a funny thing about life; if you refuse to accept anything but the best, you very often get it." ~ W. Somerset. Maughan*

> *"The impossible is possible if you are willing to be extraordinary! The only difference between extraordinary and ordinary is the word extra." ~ Annie Rose's Favourite*

Are there areas in your life where you could push yourself a little harder, but it seems like too much work? One of the reasons I found scaling to the tops of mountains so compelling is just when you felt you were pushed to your limit, you could always find what you needed if you were willing to dig a little deeper. It might not be pleasurable or enjoyable at the time, but you learn you are capable of doing more than you thought possible and the rewards were so worth it.

Terry Fox and Rick Hansen, two Canadian icons, are testaments to this philosophy. Going through physical challenges, they both pushed past the limits of what was deemed possible based on their conditions.

Terry with his Marathon of Hope where he attempted to run across Canada on a prosthetic leg running a complete marathon close to 43 times. What is truly extraordinary about that is with all of the information we have access to in today's world, the top marathoners of today only run a handful of races a year because they understand the importance of rest and allowing the body the time it needs to recuperate. Yet Terry soldiered on day after day, on one leg!

Rick Hansen with his Man in Motion tour covered 40,000 kilometres in a wheelchair! I just want to say, I assumed wheelchairs must be comfortable, but having spent time confined to one, I was so wrong.

Which is usually the case when we assume we know more than we actually do. Time for a teachable moment because the word 'assume' is composed of the letters ass, u and me.

Terry and Rick possess **extraordinary human spirit**! They had the ability to push past assumed boundaries.

Now it seems we all have things holding us back from becoming all that we could be. I know this! I am going to heal from my post stroke symptoms because being an ordinary stroke survivor holds no appeal to me whatsoever.

I am going to do extra with my recovery constantly pushing the envelope until I can re-experience the joy of writing a new song, the thrill of scoring a goal, the old man pride of being able to bench press 200 pounds, the excitement of successfully climbing through the crux while rock climbing. As it stands, where I am at presently with my recovery, the possibility of me accomplishing one of these achievements seems very remote, but I am not an ordinary stroke survivor!

What area in your life do you need to focus extra on so you can become more of the person you would like to be or have the better quality of life you were destined to enjoy? Doing extra may mean doing homework. Observe people you admire and try to emulate the traits they possess that make them extraordinary.

You see, there is method to my madness because of who I used as examples for my growth and development. I want to possess extraordinary human spirit! If you want more examples of extraordinary, just look at the performance of doctors, nurses and all other medical personnel during the pandemic pushing themselves past the point of exhaustion and thinking they were not extraordinary, but believing they were just doing what needs to be done. Taking care of yourself is important and needs to be done because we all benefit. Pick a theme, put a little extra work in and see where it takes you.

DAILY DELIGHTS!

There is awesomeness all around us and sometimes we just have to stop and smell the roses. In my humble opinion, this is one of the most important and critical practices in the entire book. Each and every single day, we enjoy good moments. It could be laughter with a friend, positive dialogue with another person, a positive result at work, an accomplishment, a compliment or kind deed given or received, a delicious meal. What I am proposing is that at the end of each day before nodding off to sleep, you take a couple of minutes to reflect on the good moments the day provided for you. At first, this may seem like work, but after a while it becomes natural and a great way to end the day. There have been countless times where I thought my day was just so so, but upon further reflection it turned out they were actually pretty good and had presented me with some extraordinary moments.

This book is focused on developing an attitude with gratitude. The whole premise being when you send thanks out into the universe for what you are grateful for, the universe rewards you with more of it in the future. Does it work? I guess you will have to try it and find out for yourself.

I believe in it because I swear it provides for better dreams and it gives you something to look forward to when you wake up the next day. Knowing this new day is going to provide you with some new delights, some expected and some unexpected. So in the words of the famous philosopher Pink, *"I'm coming up, so you better get this party started!"* And what constitutes a great party? Positive interactions with other people.

> *"I have always been delighted at the prospect of a new day, a fresh try, one more start, with perhaps a bit of magic waiting somewhere behind the morning." ~ B. Priestly*

I would tell the junior high students to focus on three to five moments. For you my new friend, I would suggest using that number as a starting point and to build on it each day. If there are days where this is hard to do, it probably means you have to initiate more. Here we go again, take action!

Be the first to dole out a compliment or suggest and organize an activity with friends. On the first Road to Recovery Tour I organized in 2018; four months to the day I suffered from my stroke. I thought it seemed like it was self-serving, but thirty people showed up and it is safe to say we all enjoyed ourselves immensely as we hiked to the top of a mountain ridge together.

I was at an emotional low point, but that day brought my spirits back to where I needed them to be to help with my recovery. It is now an annual event and consider this your invitation to join us. It keeps getting bigger and better and your presence would make that event several percentage points better which is what the gist of this book is all about. Taking action to make things better.

FORGIVE AND FORGET

I was always an eternal optimist, but after having suffered my stroke, the unthinkable happened. For the first time in my life, I began to experience the lows of depression and began to question my self worth. The crazy thing is that out of nowhere something would trigger a memory of an incident where I had done something stupid or offended someone only to further deepen the hole I was digging. And apparently, I had my fair share!

Having gone through this, I would have to say, the first person you need to forgive and forget is yourself. Like it or not, messing up is a part of the human condition and essential for us in order to grow. When you knew better, you did better! Each day is a fresh start to show the world, the new and improved you. When you are driving a vehicle, the rear view mirror only takes up a small space on your front windshield forcing you to focus on the road ahead. Without trying to sound too much like Captain Obvious, the results would be rather catastrophic if you tried to drive while only looking in the rear view mirror as so it is with life.

What happened, happened and if you didn't like the result chances are, you will get another shot at redemption. I have had to learn the importance of wiping the slate clean.

When something triggers a negative memory, I try to replace it automatically with something positive because I have had my fair share of those as well. Beating myself up was not a very productive practice, but I am working on becoming the new and better version of me. Being down only served to negatively impact those around me and they did not deserve that. As for those near and dear to you, forgive yourself and embrace the new you who is becoming the best you in the entire universe.

> "Forgive yourself for your faults and your mistakes and move on." ~ Les Brown, Motivational Speaker and Best Selling Author

> "Forget mistakes. Forget failures. Forget everything except what you are going to do right now, and do it. Today is your lucky day." ~ Will Durant

Forgiving and forgetting others for the wrong doings they have done to you may prove to be a bigger challenge. I know for me it is. Yet every spiritual book I have ever read emphatically states harbouring ill feelings towards another for their misjudgement only serves to stain your own soul preventing you from moving forward.

I had taken a group of students to an event called We Day and one of the featured speakers was Larry King; the infamous talk show host. He shared with the audience the story of his favourite guest. It was with a New York police officer who was confined to a wheelchair due to a gunshot wound.

There was this young African American boy who had a paper route and saved all his earnings so he could buy himself a new bicycle. On the first day riding around proudly on his new set of wheels, he kept getting pulled over by police officers questioning where he got the bike from. He does not know why he did this, but he had taken his older sibling's pistol and had it hidden in his trousers.

After about the tenth time, when the white policeman took him aside to ask him where he got the bike from, the young boy snapped pulling out the pistol and shooting him leaving him paralyzed. After spending months in the hospital, the officer needed to have closure to understand why this happened.

He visited the boy in detention and the young man explained the situation and his mounting anger due to the continuous harassment he had received during that day. The policeman felt embarrassment because he realized he had been guilty of racial profiling. Not only did he forgive the child, he ended up fostering him and bringing him into his own home. Today that young boy is one of New York City's finest as he too is a police officer. Wow!

How powerful is that in demonstrating how an act of forgiveness can positively impact so many? So to all those who are on my hit list, or should I say the list that rhymes with hit, you are forgiven! Even to those who didn't even realize they were on the list, you are forgiven too.

Like when someone gives a compliment hoping to get one in return, now is your opportunity to forgive me even if I haven't offended you… yet. I don't know about you , but after all that, I feel better. In fact, I would like to thank those who I offended for taking one for the team because I would like to believe I learned from my blunder and did not treat others the same way. I would like to officially offer my sincerest apologies and gratitude for helping me grow when I was weak.

> *"As I stand before the door to my freedom, I realize that if I do not leave my pain, anger and bitterness behind me, I will still be in prison. Forgiveness does not make you weak, it sets you free."* ~ Said, before Nelson Mandela left prison

Speaking of forgiveness, I feel compelled to share what I would tell my grade four classes before the first parent teacher interviews. A stressful time for many! I would tell them how excited I was to meet their parents because it would help me forgive them. It would usually take a few seconds before some of the quicker students reacted figuring it out what it was I was actually implying. Trust me the acorn does not fall far from the tree, and yes an acorn is a type of nut.

VICTOR FRANKL

Victor Frankl is another remarkable man who flies under most people's radars. He was an Austrian psychologist who dedicated his life trying to understand human behaviour. His work is second to none and I would like to share a few of his pearls of wisdom. A great deal of his motivation came from his time spent in a concentration camp during the second world war. And this is the quote I would always share with students.

> *"We who lived in the concentration camps can remember the men who walked through the huts comforting others, giving away their last piece of bread. They may have been few in number, but it proved everything can be taken from a man but the last of the human freedoms – to choose one's attitude in any given set of circumstances, to choose one's own way."* ~ Viktor Frankl

He made a bold and daring escape because he felt it was his duty to survive so he could share with the world the atrocities he witnessed so mankind would never make the same mistakes again. And so I offer these gems for you to mull over.

> *"When we are no longer able to change a situation, we are challenged to change ourselves." ~ Author Unknown*

> *"Live as if you were living a second time, and as though you had acted wrongly the first time." ~ Author Unknown*

That last quote pretty well sums up the gist of this book. It's never too late to reinvent yourself.

SHANTEL'S FAVOURITE QUOTE:

> *"YOU EITHER GET IT OR YOU DON'T." ~ Anonymous*

This is her explanation why the quote holds significance for her:

> *"It resonated with me because it's telling someone I can explain it to you but I can't understand it for you. There are people in the world who get the bigger picture how rising to the occasion brings about change in the world and change in you. And there are others who just don't get it so why waste time trying to make them understand when you could be out there improving yourself or making a positive impact."*

If you have persevered and made it this far into the book, it's probably safe to say you get it. I read once, if you are still alive it means you have unfinished business and further personal growth to pursue. My stroke was severe and left me damaged. Moving forward, I had to embrace the concept that I did not die was because I needed to learn how to cope and accept these challenges and that this book was my unfinished business.

I can't help but think, the real motivation for me sharing this quote with students was to guilt them into paying attention and to buy into what I was promoting. Once again, the focus needs to remain on us and what are we doing to grow. Rather than waste precious time chastising others in our minds for their perceived ignorance. As usual, it comes down to patience, understanding and compassion. Especially before ranting on social media or tweeting.

LIZZIE VALESQUEZ

Lizzie Valesquez, another relatively unknown personality who you would be wise to follow up on, delivered a TED talks presentation which I showed to students countless times. She first came to notoriety when someone posted a clip of her on YouTube describing her as the ugliest woman on the planet. She was born with a rare syndrome which doesn't allow her to gain weight. At birth, she weighed in at slightly over two pounds and has never surpassed sixty four pounds even though she is now an adult.

She underwent countless medical procedures and has no vision in her left eye. She grew up looking different and was verbally bullied and harassed during her entire childhood. When the ugliest woman video got posted, it went viral, eliciting all kinds of negative reactions from total strangers who reached out to her suggesting she should do the world a favour and commit suicide. Beyond reprehensible!

How did she respond to this? Of course, she was deeply hurt, but she decided to turn her situation into a positive. She adheres to the motto, *"YOU ARE IN CONTROL OF YOUR OWN LIFE!"*

She has authored four books on bullying and cyberbullying and is a powerful motivational speaker. When you check out her performance on TED talks, I am sure you will agree she is charming, delightful with a beautiful soul and the world needs more Lizzie Velasquezes and not less. We often hear how beauty is skin deep and here is proof. The next time you are being self critical of your own physical appearance, think of her and where you should really focus your attention.

MARY AND ALYSSA

Mary and Alyssa both like the quote: *"QUITTING IS THE EASIEST THING TO DO IN LIFE!"*

Funny how this quote resonated with these two, as I had the pleasure of coaching them both in basketball and the two were pistols both on and off the court. We enjoyed a lot of success so I don't understand why I obviously felt the need to keep reiterating this phrase and imprinting it on their brains.

To paraphrase a famous expression people don't realize how close they were to succeeding when they gave up. I swear I misread this, but I would always describe it this way. Apparently, Thomas Edison had tried 10,000 things to get the light bulb to function that did not work and when he was asked about his repeated failures, his reply was, "I now know ten thousand things that don't work." Which in his mind meant he was getting closer to success and I guess you can say, the rest is history.

Often in life, we opt to settle for this is good enough. On my second post stroke ascent to the top of a mountain ridge on the road to recovery tour; we were on a little plateau before the final steeper part of the hike which offered spectacular views. Not being able to prepare as much as I would have liked to, my body was beat up and the unthinkable happened.

I seriously contemplated waiting there and not going to the top. But, I am a big believer you are always rewarded when you persevere and dig deep for that little extra which you can always find. Sure enough when I made it to the top, I enjoyed one of the coolest mountain moments I have ever experienced on a mountain top.

From our vantage point, we were above a complete and spectacular rainbow. Out of the blue my daughter turned to me and exclaimed, "Look dad, Nanny is with us!" Nanny being my deceased mother who we both greatly adored.

For those who were fortunate enough to know her, if souls on the other side are capable of pulling something like this off, this would be so her. This is a moment my daughter and I treasure and will share for the rest of our lives and to think I almost bailed! How many great moments have we all missed because we took the easy way out? Interestingly enough, I was searching for quotes on the importance of optimism when I stumbled upon this quote.

> *"You'll never find a rainbow if you're looking down." ~ Charlie Chaplin*

Mind you, we did have to work for it and was it ever worth the effort.

Here is a prime example of someone who refused to give up after a myriad of setbacks and thankfully so because of their positive impact on the world and history.

- Failed in business at age 21.
- Was defeated in a legislative race at age 22.
- Failed again in business at age 24.
- Overcame the death of his sweetheart at age 26.
- Had a nervous breakdown at age 27 and would spend months lying in bed.
- Lost a congressional race at age 34.
- Lost a congressional race at age 36.
- Lost a senatorial race at age 45.
- Failed In an effort to become vice-president at age 47.
- Lost a senatorial race at age 49.
- Was elected president of the United States at age 52.

As I would read this to the class, I would ask who is thinking, "What a loser!" which would always garner some chuckles. That is, until I would mention the man's name, Abraham Lincoln!

My Americanized Canadian students would be blown away that it was him, as they realized what a positive impact he had in shaping the world as we know it today and how history would be so different had he given up. I would discuss how he was responsible for having slavery abolished and that every man is entitled to having equal rights.

It seems throughout history, those who have had to overcome the biggest struggles are the greatest achievers. So bear that in mind when you deal with your demons and battles. It is all part of the process on your expedition to personal greatness.

ROGER MURRAY

We all need a Roger Murray in our lives. Roger is a natural born leader who brings out the best in all those around him while serving as a moral compass. I greatly admire him and I am joined by a lot of other good people who share the same sentiment about him. He was my first principal who greatly influenced my professional growth and successes. He had this effect on all his staff which is the primary reason why our school was extraordinary and was renowned for its positive and warm environment. Upon my retirement, he gave me this poem in a picture frame which reflects his sageness and says it all.

Mark....

On your retirement..... I wish you enough

When I say, "I wish you enough," I want you to have a life filled with enough good things to sustain you.

I wish you enough sun to keep your attitude bright.

I wish you enough rain, to appreciate the sun more.

I wish you enough happiness to keep your spirit alive.

I wish you enough pain so that the smallest joys in life appear bigger.

I wish you enough gain to satisfy your wanting.

I wish you enough "hellos" to get through the "good byes."

"Some people come into our live and quickly go. Others come into our lives and leave footprints... and we are never the same."

You left footprints at EMHS. Miigwetch.

By Roger Murray

Miigwetch is a Canadian First Nations word meaning a deep and heartfelt thank you and miigwetch to you for taking the time to read this book.

DUSTIN'S FAVOURITE QUOTE:

"Failure is simply the opportunity to try again. This time, more intelligently." ~ Henry Ford

Let's face it. Failure is an inevitable part of life. I am pretty sure your first steps as an infant resulted in a face plant. But you got up and tried again rather than saying that's it, I tried, I failed, I give up. Resilience is a natural component of the human condition, but the trials and tribulations of life, can curb our enthusiasm to rebound after a setback.

Here is where I would give countless examples of people who initially failed only to become huge successes such as Walt Disney who was fired from his position with a newspaper for a lack of creative ideas, yes the same man responsible for the creation of Disney World, Disney Land and whatever they call the one in France.

I read somewhere how best selling author John Grisham got rejected by countless publishers which is why I elected to go the self publishing route. I would tell students if he goes to the restroom and decides to write something onto tissue paper someone would probably turn it into a movie.

Or Albert Einstein who was deemed a slow learner in his early school years and was refused entry into a polytechnic institute he applied to for not meeting the needed academic standards. Or Michael Jordan who got cut when he tried out for the senior team in high school. And for the record, that would not have happened on my watch.

> *"Failure will never overtake me if my determination to succeed is strong enough."* ~ Og Mandino

I had actually forgotten about this one which got me to thinking we need more Henry Ford quotes. A man who literally built an empire from scratch using common sense and an incredible work ethic. Enjoy these beauties!

> *You can't build a reputation on what you are going to do."* ~ Henry Ford

In other words, talk is cheap, but actions speak louder. I have always had a personal pet peeve with regards to hecklers at sporting events or at anyone for that matter who criticizes another, when they themselves have never had any experience in that field. And yes, I have extended offers to certain parents to come teach my class to show me how it's properly done.

> *"Don't find fault. Find a remedy; anybody can complain."* ~ Henry Ford

> *"Chop your own wood and it will warm you twice."* ~ Henry Ford

> *"If you always do what you've always done. You'll always get what you've always got."* ~ Henry Ford

YOGI BERRA

There is no way I could have a book with quotes in it without having some from major league baseball hall of famer, Yogi Berra. I think these are brilliant!

> *"You've got to be careful if you don't know where you're going because you might not get there."* ~ Yogi Berra

> *"Always go to other people's funerals otherwise they won't go to yours."* ~ Yogi Berra

> *"No one goes there nowadays, it's too crowded."* ~ Yogi Berra

> *"You better cut the pizza in four pieces because I'm not hungry enough to eat six."* ~ Yogi Berra

> *"When you come to a fork in the road, take it."* ~ Yogi Berra

> *"It's like déjà vu, all over again."* ~ Yogi Berra

> *"Don't let this be the greatest moment in your life!"* ~ Yogi Berra

I cannot remember who is responsible for this quote, but it definitely struck a chord with me.

I believe it came from a coach after winning a US college national football championship. After the win in the dressing room, he gathered the team around him and his victory speech was, "Congratulations gentlemen! Enjoy the win, but don't let this be the greatest day in your life!" For those who follow sports you would realize what an accomplishment of such epic proportions this was and what profound words!

One of the reasons why I was passionate about mountain climbing was that there were so many times I thought, this is the greatest mountain experience I have ever enjoyed in my life. More often than not, it would only hold the title until the next climb because you usually improved your skill set and were able to push the envelope a little further on the next one.

I always loved playing hockey and despite having enjoyed some good moments, I always felt like I still haven't played my best game and believe it is still going to happen. I actually had a song I wrote get played on a major radio station in Edmonton. That's nice but I know in heart my best song is coming even though I cannot hold a guitar right now.

The thing I like best about this quote is it provides hope for endless possibilities in the future. Who mandated your greatest moment in life has to be in one specific area? You can have a greatest moment in an athletic endeavour, a relationship, an accomplishment at work, at school, in the world of arts, as a parent, or in an area you are passionate about. All we have to do is keep pushing the envelope further and it will happen! And it can be something as subtle as this. The other day before my wife went into work, she made me an egg salad sandwich and it was the greatest egg salad sandwich I have ever tasted. It was soooo delicious!

As my recovery from my stroke is painstakingly slow, I realize my best days are behind me. But, it does not mean I still can't enjoy great days!

JOHN WOODEN

John Wooden is a legendary college basketball coach who won ten national championships in twelve years. He was a highly principled man who always conducted himself with class and dignity which was probably why he was so successful. My favourite story about him came from Bill Walton who was arguably the best collegiate basketball player at the time. They had won the national championship the year before and it was in the late sixties. With the youth revolution and rebellion of the times, Bill let his hair grow long and grew a beard clearly defying the coach's no facial hair policy. When they met to discuss it, John calmly told him, "I respect you for taking a stand for your belief in freedom of expression. That takes a lot of courage and we sure are going to miss you this year."

I would like to share some of his quotes and philosophy because we can all benefit from a moral compass.

> *"Make each day your masterpiece! ~ John Wooden"*

> *"Success is peace of mind which is a direct result of self-satisfaction in knowing you did your best to become the best you were capable of becoming." ~ John Wooden*

> *"Do not let what you cannot do interfere with what you can do." ~ John Wooden*

My most recent mountain video clearly demonstrated that.

> *"It's what you learn after you know it all that counts." ~ John Wooden*

Guilty as charged!

> *"Be more concerned with your character than your reputation." ~ John Wooden*

> *"Your character is what you are while your reputation is what others think you are." ~ John Wooden*

MATTIE STEPANEK

Matthew Stepanek was living proof there are angels who walk among us. His mother had a genetic disorder and when she gave birth to her first three children, they were all born with a rare form of muscular dystrophy and passed away as infants.

Despite having gone through this trauma, for some unknown reason, she felt compelled to try again, giving birth to Mattie who was diagnosed with the same condition. He spent a great deal of his childhood in and out of hospitals and in a wheelchair. He went through many near death experiences where he said he visited heaven. Heaven was filled with love, beauty and where he could run around and play with other children.

These experiences greatly impacted him as he tried to bring about, "On Earth as it is in heaven." He wrote a book of poems called 'Heartsongs' which were predominantly focused on peace, love and harmony.

The hospital convinced a publishing company to print the book and it sold over 50,000 copies leading to his meteoric rise to fame. Oprah fell in love with the young boy as he made several appearances on her show, sharing how life is a gift and how it is best served with a whole lot of love. Not the Led Zeppelin version. He even garnered the attention of world leaders such as Mikhail Gorbachev and became close personal friends with former US president, Jimmy Carter; always promoting a world filled with peace and cooperation.

Despite his physical impediments, he embraced life, striving to do as much as possible knowing his time here would not be long and he was always wearing a smile and was ultra positive.

I once saw him on the Larry King Show and was immediately captivated. Larry opened up the phone lines and an adult caller sharing the struggles she was currently going through and asked what advice could he offer her. A thirteen year old offering advice to an adult? As I told my students, don't go home sitting by your phone and expecting a phone call from me seeking personal guidance anytime soon. Mattie's response was sheer brilliance!

> "It's hope. You have to have hope in your heart and believe things will get better." ~ Mattie Stepanek

As I reflect on my post stroke dark times, I now trust in the validity in these words and wished I would have embraced them sooner. Without hope, we can get stuck in the quagmire of life. So trust me, no matter where you are at and whatever you are going through, believe in yourself and know better times are coming. Mattie published several other poetry books before passing just before his fourteenth birthday and here are some other insights from this wise old soul.

> "I choose to live until death, not spend the time dying until death occurs." ~ Mattie Stepanek

> "Sad things happen. They do. But we don't need to live sad forever." ~ Mattie Stepanek

> "Unity is strength... when there is teamwork and collaboration, wonderful things can be achieved." ~ Mattie Stepanek

And this is the perfect introductory quote to talk about- PLAYING FOR CHANGE. It all started when co-founder Mark Johnson was in San Francisco where he stumbled upon a street performer named Roger Ridley who was doing a killer job singing Ben E. King's, Stand By Me. Another classic song with lyrics that blend in so well with the theme of this book. He was totally enthralled with Roger's performance and curious to know how a man with such immense talent could be singing on the streets. Roger's response is pure gold and one I strive to ascribe to, "Man, I am in the joy business!" Shouldn't that be the organization we are all members of?

This triggered a spark of brilliant creativity in Mark and he teamed up with Whitney Kroenke and they formed Playing For Change, a movement created to inspire and connect the world through music. With a mobile recording studio and cameras they travelled the world and filmed musicians from different countries playing the same song who often added their own cultural slant. They meshed it all together and the results are amazing! It proves their mantra that we are all united through music is so true.

The first time I saw their finished version of, Stand By Me, I was totally blown away and it never gets old. Their productions are the absolute personification that Mattie's quote is so true. Our little blue planet is but a small island in the sea of the universe and when we pull together, we shine like mankind is supposed to. Just like every successful organization or team radiates.

I would highly recommend you check out their work. They now have a Playing For Change band as they work towards positively impacting children by helping to fund and support music and art schools. And feel free to stand by someone in their time of need because we can all use a little support from time to time said a wise lady.

Roger Ridley's line of being in the joy business got me thinking of an activity I made my grade nine students do which totally aligns with this philosophy. I cannot remember where I stole the idea from, but I made them write a letter of appreciation and thanks to an adult in the building.

I usually did this in February before report card writing started and teachers were getting a little squirrelly due to sunlight deprivation from our short winter days. I loved being the door to door delivery man. The reactions were priceless. The letters pulled at your heartstrings bringing tears to the faces to many a staff member. Some of the keener students would write multiple letters and I am sure as adults, they are in the joy business and we are not talking about the distribution of pharmaceuticals even if I do have suspicions on a few.

Paul, who received his share, placed them in a binder and always maintained it was the only thing he was taking with him when he retired. More often than not, the teacher receiving a letter had no idea he or she had that type of impact on the student and would be overcome with emotion.

We never know what lies in the future or the true significance of a kind gesture. To reiterate the importance of this activity I would tell the story of Jim who was a powerhouse teacher with a quirky sense of humour which endeared him with students. He was battling lung cancer when I visited him to drop off a student's letter.

He could not speak because of the damage done to his throat from the intense radiation treatments he had been receiving. My last memory of him was him sitting on stairs silently laughing while reading the letter over and over again. He passed soon after, but for that precious moment he was experiencing pure joy and how can you put a price tag on that?

So what is the point? Get your pens and paper out, boys and girls, because it is homework time. What I am proposing is once a year, you select a person and write them a letter explaining why it is a blessing to have them in your life. You can talk about shared memories or how they enhance your life just by being who they are. I would even offer you send it by mail to really blow their minds. I guarantee your day will be 1% better than the previous day, when they reach out to thank you. Even in writing the letter you will elicit positive vibes within yourself. Remember, this book is about taking action so I suggest to try at least once.

> "Some injuries heal more quickly if you keep moving." ~ Nick Vujicic

Nick is another amazing person whose zest for life is infectious despite the enormous struggles he has had to overcome. He was born with tetra-amelia syndrome which is characterized with the absence of arms and legs. Growing up under these conditions was a struggle for him, but through his faith in God, he found his purpose and is now an international best selling author and motivational speaker.

> "If you can't get a miracle, become one!" ~ Nick Vujicic

His book *'Life Without Limits'* is captivating, as it describes his metamorphosis from being one with a bleak future to this dynamic personality whose presentations are spellbinding.

I would always show a video of him speaking to students who were of the same ages as my classes. It always garnered powerful reactions as his humour and passion for life resonates with any viewer as you should see for yourself. My favourite segment is when he purposely falls down and struggles to get up on his one foot which he affectionately refers to as his flipper. He explains to the audience, no matter how many attempts it takes, he will rise up again, which is the lesson we all need to embrace.

Chew on this next quote and remember it comes from a man who was born with significant physical disadvantages.

> *"It's a lie to think you are not good enough. It's a lie to think you are not worth anything."~ Nick Vujicic*

Sometimes we spend too much time on our own perceived disadvantages which can hinder our growth rather than focussing on what we can achieve.

ALEX COINS A PHRASE

Now this is going to come across potentially as self-serving and stroking my ego, but it was a flattering observance and I think it has relevance to the theme in this book. Alex's mother was a colleague of mine and on their drive home one day, she was inquiring about his day. Apparently a couple of his teachers were a little more feisty than normal which is apt to happen around report card writing time. I guess he also had a Life Skills class that day to which he remarked and I quote, "You can always rely on Mister I!"

She had assumed that he was quoting me, but I had never spoken those words and what a compliment! Every time I had a Life Skills class I treated it as show time like I was performing on stage. I would amp myself up and would literally bounce off walls or put on a musical performance. What could I be relied on? Things that were in my control, like my attitude and aura that I would exude. A classic example of living in the moment. Sure there were times when I had struggles going on in my life, but they were nobody else's issues but mine and my students didn't need to bear the brunt of them. I once ran a parenting forum where the parents broke into small groups to discuss successful strategies that worked for them. One parent said something that has stuck with me to this day, whenever you see your children. You should put your stresses off to the side and act like a puppy who has not seen you for a while. That would be the personification of unconditional love and that is precisely why dog owners are crazy about their pets, so why wouldn't we be like that with our children? What a wise analogy and what I strove to embrace every time I walked into a classroom.

How would people around you start treating you if you acted like a puppy who was so happy to see them? Just don't start rubbing against them.

To be considered reliable, it involves consistency and what are the things we can be consistent with that are in our control and would endear us with others? Our work ethic, our demeanour, our attitude in trying to find positives in every situation. The last point can be challenging as I struggled to find any positives in experiencing a stroke. But, here I am two books later which probably would have never happened without it.

Here is what I propose to reinforce your consistency and dependability, write your name in the blank space provided. "YOU CAN ALWAYS RELY ON _____!"

Now let's get real here. I don't know which name you put in the blank space, but I am pretty sure it doesn't roll off your tongue quite the way Mister I does! But who cares as long as it serves its purpose. Be reliable to yourself and forge the belief you can depend on yourself to deal with whatever adversity life throws your way because it is coming.

> "If all behaviour is the result of the state we're in, we may produce different communications and behaviours when we are in resourceful states than we will when we are in unresourceful states. What creates the state we're in? The first is our internal representation." ~ Anthony Robbins

A classic example of this is something we can all relate to. What is the first thing we say when we have a morning when nothing seems to be going our way? "I AM HAVING A BAD DAY." And we often spend the rest of the day finding reasons to validate our claim and sure enough we provide ourselves with an abundance of evidence. The odds of enjoying a powerful day become greatly reduced.

What if we were to change our interpretation on those events? I would recount a personal story to clarify my point. One morning I had to have a colleague take me to work because my vehicle was in for repairs. I was running a bit late, scrambling around when I heard the doorbell. At that moment, our new puppy decided to provide me with a gift on the living room carpet. I had recently hurt my knee preventing me from working out and had gained some weight. As I bent over to scoop up the present, the button on my waist holding my pants up popped with enough force to potentially concuss someone.

Not to mention that the day was going to be the tenth indoor recess day in a row due to frigid weather which brings out the Spiderman capabilities of nine year old boys as they demonstrate the skill set of climbing classroom walls. The potential of the day was not looking rosy. I cleaned up the mess, cinched a belt around my waist and said to myself, "The day can't get any worse. If I can handle this, which I did, I can handle anything else thrown my way!"

The point is, I put myself in a state of resourcefulness rather than be a victim to the hands of fate which we all can do. Not only that, but I was able to share this story with students to make my point and it usually garnered some chuckles. It boils down to our internal representation of a situation.

> *"We always have a choice of how to represent things to ourselves."* ~ Anthony Robbins

> *The ancestor of every action is a thought."* ~ Ralph Waldo Emerson

Ralph Waldo Emerson's definition of success:

- *To laugh often and much;*
- *To win the respect of intelligent people and the affection of children; [I would always point out to the students much to their delight how intelligent people and children are used in the same phrase!]*
- *To earn the appreciation of honest critics and endure the betrayal of false friends; [Now would be a good time to reread the Forgive and Forget page.]*
- *To find the best in others;*
- *To leave the world a bit better, whether by a healthy child, a garden patch or a redeemed social condition;*
- *To know even one life has breathed easier because you have lived. This is to have succeeded.*

When Roger retired, I copied this out to him in a letter to show how much I respected and revered him because he is the embodiment of what this passage represents. Our school was known for its warm environment and he was definitely the catalyst for that. This passage represents something we should all strive for whether it be at our workplace, in our homes or whatever community we are a part of.

FAKE IT UNTIL YOU MAKE IT- is a popular phrase which is regularly used in today's world.

This can work on a daily basis. I would ask students to list the qualities happy people possessed and these are the types of things they came up with:

- They are the first to give out a greeting by acknowledging the presence of others. A regular practice that evolved into something more powerful and fulfilling for me was the giving of high fives whenever I crossed paths with a junior high student in the hallways. It started off innocuous enough, but really bloomed when a student wrote me a letter stating because she had a quiet and shy demeanour, she felt others would just leave her to herself, but in receiving a high five, it meant someone was acknowledging her existence and that she mattered. So naturally I had to take it to another level even if it annoyed the minority and the odd teacher. This simple gesture more often than not led to a positive brief interaction which always made me feel better. In fact, when there were things that were bugging me, I would wander the halls in search of students to do this with and it always put me in a better frame of mind. I believe a hello and a smile would have the same effect.
- they are quick to share a smile.
- they readily dole out compliments.

This is not rocket science and easy enough to do. So make this a regular practice and watch what happens because I know it works! Sure you may have your reasons and justification to be miserable, but to quote a comedian I once heard, *"YOU KNOW WHO CARES MORE ABOUT YOUR PROBLEMS THAN YOU? NOBODY!"*

Focus on the present interaction and create a series of positive moments that engulf your day. Your problems will always be there for when you have to deal with them, but it is always so much easier when you are in a better frame of mind.

When we enter a room, we are going to either elicit the response where our mere presence is going to enhance the energy and atmosphere or people's sphincter muscles are going to tighten. The point is walking around like you are happy not only benefits you but influences all those who you come in contact with.

HELEN KELLER

When I would talk about Helen Keller I would often be quite surprised with how few students actually knew anything about her. For those of you who may have forgotten, here is a quick recap of her story. Helen was born in 1889 and at the age of nineteen months she contracted an illness which caused her to go blind and become totally deaf. Due to such circumstances at such a young age, she was notorious for her intense fits of rage.

Her parents realizing they were in over their heads sought outside help and Annie Sullivan walked onto their lives. A remarkable woman in her own right worthy of respect and recognition for her perseverance, loyalty and dedication towards Helen.

The first thing she did was to deal with Helen's fiery bouts of rage. From there, she taught her how to sign opening a whole new world to the young child as each object had a name and its own significance. Through the use of Braille, Helen learned to read books and this became a lifelong passion for her. She became notorious for wanting to push the envelope as she enjoyed one incredible and extraordinary accomplishment after another.

For example, she learned how to talk without being able to hear sounds. In 1904, she became the first blind person to graduate with a degree from a college. She became a best selling author, international speaker who was in demand all over the world, a human rights activist as she founded the ACLU, the American Civil Liberties Union to fight for the rights of women and people with disabilities. She's one of the few who captured the hearts and imaginations of the entire planet based on her indomitable spirit and unwavering courage. A true role model for us all regardless of what generation you are associated with. With that in mind, I offer words of wisdom from Helen Keller.

"Self pity is our worst enemy and if we yield to it, we can never do anything wise in this world." ~ Helen Keller

> "Character cannot be developed in ease and quiet. Only in trial and suffering can the soul be strengthened, ambition inspired and success achieved." ~ Helen Keller

> "Optimism is the faith that leads to achievement. Nothing can be done without hope and confidence. The best and most beautiful things in the world cannot be seen or even touched, they must be felt by the heart." ~ Helen Keller

KIND WORDS

> "You have it easily in your power to increase the sum total of this world's happiness now. How? By giving a few words of sincere appreciation to someone who is lonely or discouraged. Perhaps you will forget tomorrow the kind words you say today, but the recipient may cherish them over a lifetime." ~ Dale Carnegie

This truly came to light for me when I put it out on Facebook for former students to reach out to me with their favourite quote and how it impacted them. Nicole responded to me by saying she did not have a particular quote, but she remembered something I had said to her.

She was serving an in school suspension for having been caught skipping a class. I imagine she was feeling rather sheepish about the affair as teachers would pass her in the office. Apparently, all I said to her was, "Nicole, you are better than this!"

Those words stuck with her and really struck a chord even though I have no recollection whatsoever of this interaction. Today, she is a proud momma and is seemingly doing quite well for herself. On a side note, we are all going to mess up in life and when we do, instead of beating ourselves up, we just need to remind ourselves we are better than that and move on. In the spirit in the power of words I offer;

> "Kind words can be short and easy to speak, but their echoes are truly endless." ~ Mother Teresa

> "Words have the power to destroy or heal. When words are both true and kind, they can change our world."~ Buddha

> "The real art of conversation is not only to say the right thing in the right place, but far more difficult still, to leave unsaid the wrong thing at the tempting moment." ~ Dorothy Nevill

"*STICKS AND STONES MIGHT BREAK MY BONES, BUT NAMES WILL NEVER HURT ME*" IS A LIE! All one has to do is look at the disastrous results of cyber bullying. Once something is sent out of anger or from a place of weakness, the reader can revisit the passage rekindling the pain and the hurt. We all have experienced physical discomfort, but as my father so eloquently put it into perspective when I noticed a gash on his leg, "Skin is going to fix itself, but pants I got to sew!" Physical damage will heal with time, but the same cannot be said with emotional hurt.

MICHAEL JORDAN

Being a basketball coach I had been accused on more than one occasion of using too many basketball references or analogies. I agree to a certain extent this was true because I wanted to use every opportunity to motivate my players. But in my defence, Michael Jordan stories and quotes never get old. I read once where his father told him, "*IF YOU ARE GETTING PAID TEN DOLLARS TO DO A JOB, GIVE THEM FIFTEEN DOLLARS WORTH OF LABOUR.*"

Words of wisdom to share with young teenagers soon to be entering the workforce. To which I would add, when he first joined the Chicago Bulls, he was pulled aside by a veteran telling him to ease up in practices because he had made it to the show. His reply, "And that's why you were one of the worst teams in the league last year."

As his personal accomplishments and team successes can attest to, here are some quotes from someone who has demonstrated and lived the words he speaks.

> *Some people want it to happen, some wish it would happen, and some make it happen* ~ Michael Jordan

> *"I have missed more than 9,000 shots in my career. 26 times I've been trusted to take the game winning shot and missed. I have failed over and over again in life. And that is why I am successful."* ~ Michael Jordan

From the man who is notorious for making game winning shots and still holds the record.

> *"Sometimes you need to get hit in the head to realize that you're in a fight.".* ~ Michael Jordan

And we all are in a fight to find a better way it seems.

EXPECT THE UNEXPECTED

This next story is from my young friend, Matt who is a dynamic young man sure to leave a positive footprint on the planet when his world was unexpectedly turned upside down when out of the blue he was diagnosed with diabetes.

> *"Resilience is all about being able to overcome the unexpected. Sustainability is about survival. The goal of resilience is to thrive."* ~ Jamaid Cascio

> *"Knowledge is power, and it can help you overcome any fear of the unexpected. When you learn, you gain more awareness through the process, and you know what pitfalls to look for as you get ready to transition to the next level." ~ Jay Shetty*

And as you shall soon read, no one best exemplifies that last quote than Matt!

Being diagnosed with Type 1 Diabetes completely changed my life, and my perspective. Before I was diagnosed, I had much less of an understanding or sense of my own mortality. This might sound dumb at first, but if you live the entire beginning of your life without worrying about everything you eat, taking needles every day, going blind, losing limbs, and death, it's hard to suddenly be thrown into a situation where these burdens now are either a guarantee or a strong possibility.

One day, I was the fearless hero of my own universe. Destined to succeed and live a long life. Maybe even live forever. I mean, we can't discount the fact that although it's only a slight chance, there is always talk of humans being able to live forever. I consider the possibility of 'end of life' every day because of my disease, diabetes.

When my blood sugar is high, my body is eroding from the inside out, like a sandcastle being washed away in a tide of my own blood. For those who don't know anything about diabetes, let me give you a short crash-course. My blood sugar cannot be regulated by my body naturally because the insulin-producing beta cells in my pancreas have been damaged by my immune system. In other words, the cells in my body that are meant to defend me from bacteria and viruses, attack the part of my body that produces the hormone insulin. Insulin inhibits the transfer of sugar from one's blood into one's cells. Since the beta cells that produce insulin in my body are damaged, if I don't manually give myself insulin through a needle, my cells cannot access the sugar in my blood. As a diabetic, to survive, I must control my blood sugar manually by giving myself shots of insulin.

I was seventeen when I was diagnosed with type one diabetes.

I had been trying out for the local Junior A hockey team. The tryout process for Junior A hockey lasts for months. Over about three months I noticed I had lost thirty pounds. I felt fast on the ice because of it, but I would easily be pushed off the puck. I began to notice my muscles would become exhausted easier than before. I ended up getting released back to my Midget AAA team about a week before the season started.

One practice, we were doing full ice battles; it's one of those drills where the coach just yells out a number and that many guys jump off each bench to battle, and in my case it was one-on-one. The coach called out; "ONE!" and I knew I had to show my dominance in this battle, as I was a veteran on the team returning for my last season of minor hockey. The coach chipped a puck in, and I skated as hard as I could to get to the puck first, and then position my body between him and the puck so that I could control it while evading him. I skated around him wide and was flying in on net when I noticed my vision beginning to fail me.

During this drill I had overexerted myself in my excitement. A white tide closed in on the center of my vision. I had to leave the ice because I was extremely lightheaded, my ears were ringing and all my muscles were cramping up. I was mad and felt as if my body had failed me, or rather, I had failed my body by not training harder in the summer. The trainer met me in the dressing room along with my dad, and after discussing my symptoms he decided it would be best if I went to the hospital.

I walked into the hospital assessment office and after taking my vitals, the nurse checked my blood sugar. The nurse informed me thirty millimoles per liter popped up and this was way too high!

After more testing and being hooked up to an intravenous flush all night, the doctor finally came in to consult with us. The doctor said to my dad, "Mr. Sparrow based on our test results I'm 100 percent sure your son has diabetes."

"What does that mean dad? Am I going to be alright?" I remember him reassuring me I would be okay at the time, but later would admit that he was in a state of shock. I wondered

if people were going to make fun of me. To add salt to the wound, after being released from the first Junior A team I was trying out for already, being off my skates for the next two weeks minimum would prevent me from attempting to go to another camp.

I live with one roommate on Vancouver Island and I'm studying business at the University of Victoria. I've travelled to different continents, graduated with my friends, continued my hockey career, and moved away for university all as a type one diabetic. I have not allowed my disease to stop me from enjoying any aspects of my life, in fact, people are often surprised when I tell them that I'm a diabetic at all. I get questions; How can you play sports? Isn't it unsafe for you to live on your own? How do you take care of yourself? I never really know how to answer. 'I just do it' is about the best I can muster.

Obviously, things are a little more difficult for me than a normal person, but things could also be a lot worse, and I'm extremely grateful for the privileges that I have. I couldn't ask for better support from my friends and family. Now that this has become my new normal, it seems foreign to consider things being different. Some days I do find myself looking for distractions from the stress of diabetes, and I feel this is normal, given the demanding nature of the disease. As a diabetic it's my life mission to control and correct my blood sugar. I take pride in my ability to control my blood sugar, so when it's out of control it becomes frustrating fast. Waking up in the morning, feeling gross, does not make for a good day. However, I have found the innate maintenance diabetes requires is not all bad.

In learning to better regulate my nutrition and activity, I've gained the knowledge to fine-tune my body. In adhering to this regulation, I've gained willpower. My multiple daily injections constantly remind me that pain is temporary and necessary for growth. Often people are offended or confused when I don't show interest in the same foods as them because I value how I feel post-meal much more than I value taste or craving when making a decision on food. Diabetes has made me hyper aware of my circumstances, and therefore I revere all decisions in my life with a greater intensity and severity, for better or for worse. Although it is a burden, diabetes is manageable. If I persevere and find ways to deal with the day-to-day challenges, I can live a normal life. Activity, carbohydrate counting, in addition to proper insulin dosages will reduce fluctuations in my blood sugar. New discoveries in the field have made large improvements to treatment in recent years, and there is potential for a cure in the future. Help and support from my family, friends, and doctors, as well as an understanding of my disease puts me in an opportunistic position regarding health. Both the short and long-term effects of diabetes are up to me to prevent. The new outlook on life that came with my chronic illness gives me a more unique perspective and motivates me to defy the statistics and stereotypes. I won't let my disease define me. Let the balancing act continue, I wouldn't have it any other way.

By Matt Sparrow

THE GRADE NINE MOUNTAIN HIKE

I always felt so empowered standing on top of a mountain. Through the physical discomfort and grind to the euphoria from achieving your goal with spectacular views until you realize you still have to make your way back down. I referenced this endeavour so often, I knew I had to find a way for the students to experience it for themselves. There is a hike in Jasper National Park which is safe and it was long enough to physically tax the students, rewarding them with majestic views for their hard earned efforts.

This activity provided me with new material every year that I could and would use with future classes. Of course, the obvious was demonstrating the power of working together as a team as it was arranged so we would all summit at roughly the same time and how we are rewarded when we push ourselves. Here are some examples of some memorable moments.

NICK'S SELFLESSNESS

One year when we made it to the top, we got pelted with hail and rain, making the conditions cold, wet and miserable. When we made it back into the trees, I came across a group of students. There was a girl sitting off to the side who started to cry because she was cold and uncomfortable. Nick, who had a degenerative heart condition which stunted his growth as he probably didn't weigh more than seventy pounds at the time, was wearing shorts, a tee-shirt and a light jacket. He promptly takes off his jacket handing it to the upset girl and says, "Here, you need this more than I do." Even though she had on several layers of clothing!

I would retell this story every year and it would replay itself time and time again with students verbally encouraging those who might have been struggling or even carrying their backpacks for them. A lesson for us all as nothing garners respect quicker than the act of putting someone else's needs ahead of our own.

FRAN COMES THROUGH

This occurred on the very first trip and I was a bit antsy wanting to make sure everything went smoothly. We were about an hour and a half away from the starting point on the drive up, when we ran into what seemed like a torrential rain storm. Not knowing what else to do, I started praying to Fran to help me out.

She had just recently succumbed to cancer and she was the most maternal grade nine teacher. Her students loved her dearly because they knew she cared and would fight for them to be successful. Knowing her passion. I figured if anyone could pull some strings for us in heaven, it would be her. Did I mention she was a feisty Italian from New Jersey who could have put Tony Soprano in his place?

Needless to say as we approached, the conditions began to improve to the point when we got off the bus, there was only a slight drizzle. We successfully made it to the top and got ready to pose for a group photo shot when a boy said look up.

Directly over our heads was a circular patch of blue sky, the first we would see that day! So what is the point of sharing this story? I think it's cool and it can be reassuring to believe there is an Other Side who is willing to help us in our time of need.

GLYNNIS PERSEVERES

Glynnis was a young girl who was not the most physically fit student. In fact, I had teachers question whether I was going to take her on the trip and whether that was a wise decision. But, I knew Glynnis was a battler and had a big heart who would rise to the occasion when she challenged herself.

I would warn students they may be stiff and sore the following day, feeling like rigor mortis was setting in. It took her some time to make it back down and with the smile of a champion, these were the first words she said to me, "You lied Mister I, you said we wouldn't feel the effects of rigor mortis until tomorrow!" I have experienced the physical pain which she was going through, but I am pretty sure she would tell you it was well worth it. Sometimes we need to go through some discomfort to enjoy a great moment but such is life.

BETTER THAN AWESOME

The question is, how can we create scenarios that are better than awesome? On one sun filled trip, a student said, "Mr. I, this is just awesome!" And I responded with, "No, this is better than awesome!" She must have thought I was being melodramatic because she challenged me by asking, "How can something be better than awesome?" I am on the clock and getting paid right now!" was my reply and checkmate for the win. Creating moments that are better than awesome takes time and effort but the results are pure gold!

'Nine Lessons I Learned From My Father' is a wonderful book written by Dr. Murray Howe about his father, the legendary Gordie Howe. It would seem as amazing as he was on the ice, he was even more amazing as a person. I cannot think of a greater honour for a parent than to have your own child pay tribute to you in such a heartfelt and sincere manner. It is a wonderful read and I am just going to list the chapter titles as there is no way I could do the book justice if I tried to describe it in any type of depth, but the headings give a fairly obvious idea what this legend of a man promoted and modelled.

Chapter 1- Live Honourably

Chapter 2- Live Generously

Chapter 3- Play Hard, But Have Fun

Chapter 4- Patience, Patience, Patience

Chapter 5- Live Selflessly

Chapter 6- Be Humble

Chapter 7- Be Tough

Chapter 8- Stay Positive

Chapter 9- Friends and Family Are Like Gold, Treasure Them

I guess the moral of the story would be you don't need to be blessed with God given talents to become a superstar in life! Now I never had the privilege and honour of meeting Gordie Howe, but I know a man who greatly resembles him and who was definitely cut from the same cloth. Yes, that would be Roger. Here is an interesting challenge. Try to connect a famous personality with someone you know. And even though we both have similar hair styles and sport moustaches, please do not confuse my intentions with trying to be Dr. Phil!

JOE AND JOSEPHINE IVANCIC

This just seems like the ideal time to speak of my mother and father, **WONDERFUL PARENTS** with solid morals, who were hard working and placed family above all else. They grew up in a small majestic village nestled in the mountains in what is now known as Slovenia.

I have visited their birth place on several occasions and I am always awestruck by the beauty of the landscape and the warmth of the people. In my humble opinion, it is like a slice of heaven on Earth. They grew up in impoverished conditions and wanted better for their children. During this time period, attempting to flee the country formerly known as Yugoslavia could result in being thrown in jail or even death.

Despite the possible negative ramifications, they took the risk immigrating to Canada to try and start a new life without knowing a word of English or French and possessing no real employable skill set other than determination and a hardy work ethic.

Initially, times were tough to say the least, taking on one difficult job after another from working in bush camps in the frigid Northern Ontario winters to working underground in the gold mines in Timmins. They toiled and saved until they could afford a small house which would serve as the home for the five of us.

We didn't have a lot of extra amenities or lavish vacations, but we had enough to experience, and here I speak for my brother and sister, a wonderful upbringing that provided us with warm memories filled with plenty of laughter and enough support and opportunity that we could all land on our feet as adults.

Not having much of an opportunity in their homeland, my parents valued education as my father would constantly remind us, "ANYONE CAN WORK SQUARE SHOVEL, BUT YOU GOT EDUCATION, YOU GOT A CHOICE." I would use this line repeatedly with students in an attempt to motivate them on the importance of trying their best in school; by telling them if I wanted to quit teaching tomorrow and get a job as a roofer I could go and seek employment for it. But if a roofer woke up one morning and decided he wanted to be a teacher, the situation would be much more complex and difficult. The lesson being, to do your best to improve yourself in the moment so you can have options and opportunities down the road.

My father fell in love with the game of hockey, attending as many of our games as possible, but he never critiqued us or overstepped his boundaries as a hockey dad. Which I am so grateful for. But he possessed that gift to say the right thing to make you think and smile at the same time.

For example, in my teen years when I was unsuccessfully trying to be a tough guy on the ice, my dad would shake his head and say "Mark, I never see anybody score goals from the penalty box!

Logic without passing judgement, as he was in a lot of cases with us. Mind you, my brother and sister gave him plenty of opportunities to dispense his philosophical wisdom which we still share and laugh about to this day.

My parents were staunch supporters of their children and would do whatever was needed to give them any advantage. A valuable lesson for all of us in our seemingly what's in it for me times. Putting family first is such a powerful priority to possess. Of course, except when your children come home and complain about their teachers.

Now my mom had her moments as well, often telling me for reasons unbeknownst to me, "Mark you gots a funny brains!"

After my stroke, there is now scientific evidence to support her claim! On those rare occasions where I would cross the line, she would boldly state, "Mark I'm going to wrinkle your neck!" I was never quite sure what that entailed but it did serve to get me to rethink what it was I was doing at the time to get her to say those words to me.

When I think of my parents, I think of love and perseverance. What will our children think of us? Both of my parents have passed away sadly and I delivered eulogies for both of them. It was easy in one regard in that I had so much material to draw from to demonstrate my gratitude. What is the legacy we are leaving behind in the eyes of others? I would tell the students on my tombstone and at my funeral, I want the sentiment to be, "Mark was a good friend!" Only time will tell whether or not I achieve my goal, but it is such a lofty aspiration that has served me well and naturally makes my life one percent better.

LEO BUSCAGLIA

Leo was also known as Dr. Love which all started in the late sixties when he was a professor in the Department of Special Education at the University of California. A student in one of his classes committed suicide and this event greatly impacted him because she was a young woman who he enjoyed having in his class. He felt compelled to do something so he offered a course simply called Love examining how love can conquer so many of the societal woes in our world.

Ironically, this was a major premise in my Life Skills class even though I was providing a Dollar Store version and looking back I should have used more of his material. He did a speech on PBS television which really struck a chord with the viewers, propelling him in becoming a best selling author and a much sought out motivational speaker.

I read his book, 'Living, Loving and Learning', in my early twenties and it made a huge impression on me. This book inspired me to read more of this type of literature, which I highly encourage you to do. In honour of the late great Leo Buscaglia I offer these messages on the importance of a world filled with love.

> *"Too often we underestimate the power of a touch, a smile, a kind word, a listening ear, an honest compliment, an act of caring, all which have the potential to turn a life around."* ~ Leo Buscaglia

He was infamous for his giving of hugs.

> *"Love always creates, it never destroys. In lies man's only promise."* ~ Leo Buscaglia

> "I believe that you control your own destiny, that you can be what you want to be. You can also stop and say, "No, I won't behave this way anymore, I'm lonely and I want people around me, maybe I have to change my methods of behaving. And then you do it." ~ Leo Buscaglia

> "I have a very strong feeling the opposite of love is not hate – it's apathy. It's not giving a damn." ~ Leo Buscaglia

THE EFFECTS OF KIND DEEDS

I read somewhere how when you receive a kind act, the brain releases dopamine and serotonin, chemicals responsible for making us feel joy. Not only that, but the giver of the kind gesture receives the same neural response. I would tell students, if you are unhappy and not feeling good about yourself to do a kind deed for another person.

In fact it gets better, even an observer of the act is similarly impacted. Hence, one good deed in a crowded area can create a domino effect and improve the day of dozens of people. Kindness is contagious! The positive boost lasts for three to four minutes, so acts of kindness have to be repeated in order to maintain, a 'Helper's High'. Science has proven after helping others, you are calmer, more positive and energetic and less depressed. As Dr. Ishak describes it, "*We build better selves and better communities through kindness.*"

> "People who volunteer tend to experience fewer aches and pains. Giving help to others protects overall health twice as much as aspirin protects against heart disease. People 55 and older have an impressive 44% lower likelihood of dying early and that's after sifting out other contributing factors, including physical health, exercise, gender, habits like smoking, marital status and many more. This is a stronger effect than exercising four times a week or going to church." ~ Christine Carter, Author, 'Raising Happiness in Pursuit of Joyful Kids and Happier Parents'

So after writing that quote, I can't help but think, how in the @#$%! Did I end up having a stroke? But, the bottom line is I am still here!!

UNITY IN THE COMMUNITY

After sharing that with students, I felt compelled to find a way to put that information to use. With the help of student teacher Rachel who came up with the cool name of Unity in the Community, we devised a volunteer program for the grade eight students. I had the volunteer coordinator from our community speak to them explaining how to go about offering their services and where the best and most needed opportunities were.

I prepared volunteer sheets that had to be signed by an adult. The students had to do twelve hours of community service in exchange for the opportunity to participate in a twelve hour wake-a-thon out of the school. It was such a win-win for everybody involved. Our school families got free tutors to help their children, our basketball program had an endless supply of volunteers to work the score clock, and a tremendous amount of help was provided in the community, mind you I suspect the pet shelters benefitted the most. Not only did this boost the morale and self confidence of the students, but it was something they could always put on a resume and some even managed to get employment where they volunteered. On a yearly basis, they would provide close to a thousand hours of service.

But let's not kid ourselves, the wake-a-thon was the real inspiration for them which could have never happened without the great support of my colleagues and administration. I had a schedule of activities starting off with Alamo dodge ball as it always ended up with a handful of teachers facing off against fifty plus students much to their delight as it always resulted in pure carnage for them.

We would have midnight pizza followed by karaoke and at about four or five in the morning, we would play everyone's favourite, Hunt The Student. Students were given ten minutes to hide in the building, then sleep deprived crazed teachers hell bent on revenge armed with water guns and super soakers would scour the hallways looking for victims. This whole evening was a lot of effort, but it enhanced the school vibe in the building and created a better rapport between staff and students which is always the by-product of sharing laughs with others.

Today's lesson and challenge is finding unique ways to create laughter and memories with others! And sometimes it may require more effort and energy than you would like to give but it always pays off!

THE EGM SOCIETY

So often we look at outside sources for powerful human interest stories when they exist all around us. The Evan Grykuliak Memorial (EGM) Society is a bullying prevention organization based out of Edmonton. Unfortunately, it was founded by Evan's friends after a most tragic set of events.

One of my favourite quotes is, *"IT'S NOT WHAT HAPPENS TO YOU THAT COUNTS, BUT MORE IMPORTANTLY, HOW YOU CHOOSE TO DEAL WITH IT!"* Which is why this is such a powerful story that needs to be shared.

It all started off innocently enough, as friends and family gathered together at a rented hall to celebrate this dynamic and charismatic young man's seventeenth birthday party. All was running smoothly until some uninvited young men tried to crash the party. Not wanting any problems, Evan greeted them at the door and calmly explained it was a private affair.

For reasons that are still unknown to this day, one of the intruders pulled out a knife and proceeded to stab him, taking the popular young man's life. As one can imagine, this greatly traumatized those who were there and witnessed the event considering they were all teenagers and still in high school.

A fundraiser was held out of their school and not knowing what to do, a group of friends decided the best way to honour their friend who was very community-minded would be to set up a non-profit organization to work with schools and students to work towards preventing similar incidents from reoccurring.

Being how Evan was an avid soccer player, one of their first initiatives was setting up a soccer tournament for youths, Kicking Out Bullying, using the funds that were generated to help schools with programs such as The Leader In Me. What an example of building something positive out of something horrific.

I had a student who was insistent I meet with his uncle who was president of the organization at the time thinking we would be a good fit because we shared like-minded philosophies.

We arranged to meet, to get to know about each other. We were both into developing programs for youths so the meeting was rather exciting as we shared our passions and visions of possibilities. As we were sitting there, a thought entered my head, "Now this is the type of young man my daughter needs to meet." When I got home, I think I may have pumped too much air in Dan's tires and under the guise of be careful what you wish for, the two are now officially engaged to get married. Now this overprotective father has to spend the rest of his days begrudgingly being labelled as a great wingman. Sheesh!

THE POWER IN ME

I attended and brought students to many We Day productions which were truly epic and powerful events that focused on getting youths to do volunteer service both globally and locally.

My favourite parts were the speakers who would try to motivate and inspire the audience with their stories and the world class musical talent who would provide amazing entertainment.

It was always a great day and my students would be so excited. I always felt it catered to the movers and shakers who already got it but what about the kids who were struggling? So together with the EGM Society, we developed The Power In Me. A day filled with incredible local musical talent, minus the fact I squeezed myself into the line up, and people who shared stories about how they overcame major obstacles in their lives. They are all over the place and have so much to offer. One such person who is going to share her take of the event was Kelly Falardeau who is a burn survivor and whose story is so impactful we had her speak at three of the four events we hosted. Kelly is a multiple TEDx speaker and 7x Best-Selling Author whose passion is to help people to make lemonade out of lemons. Here's what she wrote me:

Dear Mark –

I still remember the first Power in Me event when you told me I was going first. I said "What?! Are you kidding me? Why first?"

And your comment was, "Because you have a powerful story and the kids need to hear it and I want to start the event off with a bang!" I was so crazy nervous, but I pushed myself and did it anyways! That's what I was taught, feel the fear and do it anyways.

One of the reasons I love you is because of how humble you are and how you can see the power in someone's story; even when they don't see it themselves. Like me.

You see, when I spoke at schools and the teens were all quiet, I thought they were bored, but no, that meant they were engaged, present and paying attention. I honestly didn't realize how powerful my story was until I saw the video footage from my speech and saw some of the girls crying. I still remember you telling me after I spoke to your Life Skills Class; "Kelly tell them more about your story, it's so powerful."

In the third Power in Me speech I did, I was talking about the mirror (and even had a full length mirror on stage) and how we think it talks to us, but it doesn't. And when we look in the mirror, we let it tell us mean and negative things about ourselves. And as a result, that's how we develop such negative self-talk.

In this presentation, I was able to show the teens the mirror DOESN'T TALK. We learned from watching Snow White when the evil Queen talked to the mirror, that it talked. But the mirror doesn't talk and it's our own thoughts that are harming us and we need to change those negative thoughts to positive ones.

What was so cool is, that presentation made it into a documentary about my life story called 'Still Beautiful' and you and Ben sang the Still Beautiful song for 3000 teens to hear and they loved it. And now that documentary has aired across Canada, multiple times.

Every time I spoke at the Power in Me event, I came away feeling so empowered and grateful for the opportunity to share my message with so many teens. Thank you Mark for this huge opportunity to empower the teens and myself.

Kelly Falardeau

> "Dreams are meant to be found, not tucked away in Dreamland." ~ Kelly Falardeau

THE MEDITATIVE ALPHABET

Probably more from age, but since my stroke I have difficulties with sleep. So I practice something which I call the Meditative Alphabet to distract my thoughts and to redirect my focus to something more positive and beneficial. Starting with A, I associate a series of words that begin with a specific letter and work my way down. The list is constantly changing, as I will now provide examples.

A-	AFFECTION FOR ALL ESPECIALLY ANN MARIE • ASPIRE TO ACHIEVE • AMAZING ADVENTURES AWAIT • AN ATTITUDE OF APPRECIATION
B	BLISSFUL BEAUTY TO BEHOLD • BEST BUDDIES- and here I pray for wellness for friends who are ill or struggling or simply give thanks to all the wonderful people in my life
C-	CONQUER THIS CHALLENGE LIKE A CHAMPION WITH COURAGE, COMMITMENT, CONFIDENCE AND CHARACTER
D-	DAILY DELIGHTS- time to reflect on the blessings of the day • DEFY THE ODDS • DEFEAT THE DEMONS • DELIVER ON THE DREAM
E-	ELEVATE THE ENDURANCE, THE EXERCISE, THE ENTHUSIASM • ENHANCE THE EDUCATION, THE ENCOURAGEMENT, THE EMPOWERMENT • ENJOY EACH DAY • EVOLVE OR DISSOLVE
F-	FORGIVE AND FORGET • FOCUS ON FAMILY, FRIENDS, FITNESS AND FUN
G-	FAITH IN GOD'S GRACE, GUIDANCE, GRAND PLAN, GREATNESS • GIVING • GRATITUDE • GUITAR PLAYING
H-	HEAL • HEALTHINESS LEADS TO HAPPINESS • HARMONY FOR HUMANITY • HOCKEY PLAYING • HOPE

I-	BE AN INCREDIBLE, INSPIRING IVY, Ivy is my childhood nickname
J-	JOY • JUSTIFICATION FOR JUBILATION- because I will recover • JOSIE- my dog named in honour of my mother • JOE AND JOSEPHINE- I have to give a shout out to my parents • JACKED- because I want to lift weights again
K-	KINDNESS • KNOWLEDGE
L-	LAUGHTER • LIVING, LOVING, LEARNING, LIFE
M-	MAGNIFICENT, MARVELOUS MEGAN- my daughter • MEDICAL MIRACLE MARK • MOUNTAINS
N-	NONSTOP NATALIE- my sister who refuses to lose hope with me • NEVER GIVE UP NANNY, my mother who was always ready to fight the good fight with heart and soul
O-	OVERCOME THIS OBSTACLE WITH OPTIMISM AND TREAT IT AS AN OPPORTUNITY TO OVER ACHIEVE
P-	TURN POSSIBILITIES INTO PROBABILITIES • POWER • PERSEVERANCE • PEAK PERFORMANCE • POSITIVE PERCEPTION AND PORTRAYAL
Q-	QUIT THE SQUAWKING- at the end of my career, I felt I was being disrespected and unfairly treated. I started to complain which only served to make things worse as it negatively impacted my career and self image.
R-	RECOVERY • REHABILITATE • REDEMPTION

	- RESURRECTION- after my stroke, I thought part of the old me had died and I was so wrong! - THE RETURN OF MR. I
S-	SPECTACULAR SUCCESS STORY
T-	TEMPORARY TROUBLES - TREMENDOUS TRIUMPH
U-	UNDERSTANDING - UNLOCKING THE MYSTERIES OF THE UNIVERSE- I am a spiritual man and believe there is a positive loving creator ready to help us on our journey
V-	VICTORY - VITALITY
W-	WORDS OF WISDOM - WITNESS THE WORLD'S WONDERS
X-	EXPECT EXTREME EXPERIENCES - EXCEL - EXCELLENCE
Y-	YOUTHFULNESS
Z-	ZEN - ZERO IN ON GETTING INTO THE ZONE - ZEST, ZIP AND ZEAL - ZIPPETY DO DAH ZIPPETY DAY, MY OH MY WHAT A WONDERFUL DAY!

So much for a brief overview. The best part about this is coming up with words that you need to personally focus on as it can always grow and be enhanced. I find it redirects my focus in a more positive direction and helps me fall back to sleep after my post midnight pee breaks. I am sixty!

OPINIONS

I find it amazing the role opinions now play in the modern world due to the influx of social media. Somehow people's opinions have more relevance and influence in our lives like never before. And after saying that, my opinion is that it is not a coincidence mental health issues are skyrocketing and affecting people in such negative ways.

There has never been a time where the need to fit in or appear as sensational has been greater, adding more stress in our lives, making us lose sight of the only person who we need to truly impress, OURSELVES! Do that and everything else falls into place.

I could be wrong, but I think this quote came from former NBA hall of famer Charles Barkley who sums it up best:

> *"Opinions are like butts, everybody has one and they predominantly stink! Especially when they make sounds!"~ Charles Barkley*

Keep that in mind when someone is trying to display superior intellect in expressing their views on world matters or even with what they think of you. It is insanity to let other people drag you into their quagmire. When people's opinions of you are negative, it is usually an indication they are not in a very good place and they need to bring you down in a feeble attempt to make themselves feel better. Yet our self-worth can be predicated on how many likes or dislikes we receive from strangers when we post something. Guilty, as I find myself posting more stuff on social media.

Which is why I love the phrase, *"I AGREE TO DISAGREE!"* You are entitled to your opinion, but by no means am I obligated to have to subscribe to it. Having said that, if your opinion is that this book has merit and value, bamma ramma slamma jamma! If not, take it up with Souchie, keeping in mind his sign about the rising cost of ammunition!

"FIRST YOU SUFFER AND THEN YOU SUMMIT" is a slogan from North Face and on a tee shirt I received as a gift from my daughter. I cannot help but think, how this is such a great

metaphor for life. It is an inevitable fact of life we are going to have to deal with some form of suffering. As much as we try, we cannot escape the reality we will experience pain in our lives, as it is all part and parcel of the human condition.

And every time we deal with pain, we grow stronger which enables us to deal with the next difficulty thrown our way. Self-belief you are strong enough to handle your woes can go a long way because inevitably it has to be you who puts in the work. What we can control, is how we choose to deal with it. Here lies the challenge. Before I suffered my stroke, I believed I was strong enough to deal with anything life could throw my way, but in an instant I was put to the test and initially, I did not do well.

It is so easy to administer what people consider to be sage advice from a distance and a point of logic, but it can be irrelevant until you arrive at a stable place emotionally. I now have a new respect and appreciation for those dealing with mental health issues. The struggles are real and can be so debilitating. People with good intentions will tell you what you should do, which can be much easier said than done. The best I can offer to anyone struggling, is you know you best.

There has to come a point when you have strength and you develop your plan of action. For me, it involved working on my fitness, going out in public so I could interact with others, going back to the gym, reading and writing with a focus on inspiration and positivity. It will take time because I am not totally back to my old self but there was a sign I read at my rehabilitation facility which I constantly remind myself. *"I AM NOT BACK YET BUT I AM CLOSER TODAY THAN I WAS YESTERDAY."* And just like hiking to the top of a mountain, it is taking steps and moving upwards. It may be trying and discomforting, but the end reward is always worth the effort.

A FAVOURITE MOVIE QUOTE

After the previous page, I am sure one would assume, I would be quoting Arnold from his role as the Terminator with, "I'LL BE BACK!" And I will be, but once again without being overly redundant, that is not my all-time favourite movie quote.

It comes from a little known movie called Stigmata, where a woman inexplicably develops the wounds of Christ on her body like he had at the crucifixion. She has wounds on her wrists and it is assumed she has attempted suicide. So she is taken to an institution for evaluation.

As she attempts to defend her innocence, she cites this line which I think is pure brilliance, *"I WOULD NEVER HURT MYSELF! JUST ASK ANYONE WHO KNOWS ME. I LOVE BEING ME!"*

I love being me means you know you are not perfect, but you are enjoying your life and you are in the midst of one amazing expedition which is how I felt pre-stroke. My life was very active in a variety of ways and I was having a lot of fun doing things with good people. It just means I now have to figure out what the next expedition is going to look like because it will be different, which does not mean it has to be less fun.

From Men in Black 3, I offer, *"MIRACLES ARE EVENTS THAT ARE SEEMINGLY IMPOSSIBLE THAT HAPPEN ANYWAYS!"*

One does not have to search far and wide to realize this is a daily occurrence which means anything is possible. This should foster unlimited hope a miracle can happen to us if we open up our hearts to the notion. I am banking on it and I will keep you posted with how that works out for me.

When we go through trauma, we need to believe we are worthy and capable of a miracle which is the first step in making it come to fruition. I know by talking and writing about my situation, I believe, why not me? I mean I have already achieved so much strength emotionally and physically which has led me to accomplish and enjoy some pretty cool moments. With more to come! So why can't you heal as well? Because somebody has to lead the way with whatever you are going through.

"Don't go around saying the world owes you something; the world was here first, it owes you nothing." ~ Author Unknown

This quote evokes memories of a young adolescent I saw who yipped at their parent stating, "If you don't want to give me what I want to make me happy, then why did you even have me?" "Because we didn't know it was you we were going to have."

Now I don't profess to be able to read people's minds, but based on the expression on the parent's face, I am pretty sure what they were thinking was something like, "That's a really good question!"

Based on my many years of interactions with children, it would seem a surefire recipe for frustration and disappointment would be to start with a heaping mound of a sense of entitlement adding regular amounts of expectations others will be there to meet your needs. Expecting people will come running to your rescue in your time of need seems like a good formula for losing friends. Now don't get me wrong because having people by your side in times of duress is a powerful thing, but depending on it seems to be an exercise in futility. And now I would quote lyrics from an Avril Lavigne song, *"Why should I care? Because you weren't there when I was scared. And I was so alone."*

"Boo hoo hoo!" Whatever it is you are going through, you got this because remember, you can always rely on.... your name.... To handle and deal with a situation. This has to be true because it is written in a book somewhere.

And here is a scenario I would describe. If it was twenty below and raining and my car broke down leaving me stranded by the side of the road. Twenty below and raining? Really? I used to like messing with their minds from time to time to see if they were actually paying attention. And if a student drove by with their family, waving at me, but not stopping. I would tell the students my response would be to smile and wave back because I would be able to solve my problem on my own somehow.

Now, I would add if the roles were reversed, my expectation of me would be to stop to see if I could help with the situation even if I possess zero and I mean zero mechanical abilities. The point being, instead of placing expectations on how others should be there for us, we should be focusing on ourselves and how we can improve situations for others. It is so much more productive. I have received sympathy for my current state of affairs and it has never once helped me improve the range of motion with my left arm!

LIFE'S LITTLE INSTRUCTIONS

I would receive material from different sources such as a sheet like this one which had about fifty quick pointers on how to best live your life. I am not going to write them all but would rather focus on key tips I personally thought deserved emphasis.

- BE KINDER THAN NECESSARY
- COMPLIMENT THREE PEOPLE EVERY DAY
- COMMIT YOURSELF TO CONSTANT IMPROVEMENTS
- BE THE MOST POSITIVE AND ENTHUSIASTIC PERSON YOU KNOW. When I would challenge students with this, some would respond with, "We can't compete against you." Well, after having suffered a stroke, in my eyes, it is quite apparent I no longer possess the championship belt. But, I am determined to regain the crown! So if you have aspirations in claiming the title, consider this an open ended invitation to bring it on and good luck because I want my belt back!
- THINK BIG THOUGHTS BUT RELISH SMALL PLEASURES To me this signifies living in the moment and enjoying the simple things in life.
- WAVE AT KIDS ON SCHOOL BUSSES And you don't have to be as exuberant as teachers on the last day of the school year, speaking from experience.
- LOOK PEOPLE IN THE EYE AND HAVE A FIRM HANDSHAKE. Nothing creepier than the dead fish handshake. I would compare the two with students and the latter would always freak them out.

> "As a well used day brings happy sleep so does a well used life bring happy death."-Leonardo Da Vinci.

Happy Death? You are probably thinking this is an oxymoron and a contradiction of terms. Like calling someone pretty ugly or as Souchie wants me to emphasize the fact he considers himself to be an educated redneck and with all due respect, I had no idea such a thing existed.

The point is, we all have experienced days where they are just filled with activity and positive interactions. A day so full leaving us exhausted and upon reflection of it as our head hits the pillow, it feels like we literally fall asleep with a smile on our face. There will come a time for most people where we find ourselves in the situation where we reflect on our lives or have our lives flash before our eyes and it better be something worth watching. Time for some lyrics. From Bon Jovi, *"I'M GOING TO LIVE WHEN I'M ALIVE AND SLEEP WHEN I'M DEAD!"*

This has been a challenge for me, but I have learned to aspire and dream again which refuels my engine to focus on new positive possibilities. I found I had to redirect my focus towards new accomplishments even though they were not on the terms I wanted them to be.

THE WATER CRYSTAL EXPERIMENT AND THE POWER OF INTENTIONS

There is a controversial video I would share with students about how the effects of words and intentions can affect the molecular structure of water crystals. Masaru Emoto conducted these experiments where he would label glasses of water with positive phrasing such as I love you or references to kindness, peace and harmony and while attaching the messages he would focus on those sentiments.

He also did the same with negative messaging such as you are ugly and I hate you. He froze the water and with the use of a special microscope he observed how the water crystals formed. Interestingly enough the results were remarkably different. The water crystals in the positive messaged containers developed into uniquely structured and beautiful formations in comparison with the chaotic and discoloured formations with the negative intentions.

As I mentioned, the results are controversial, but check it out for yourself. It got the attention of the students and that was the purpose of showing the video. When you consider how much of the human body is comprised of water, this demonstrates the power of intention and human consciousness. There is countless medical evidence which supports how angry people are more prone to heart disease and cancer. This just goes to further support the importance of controlling our thought patterns and how we choose to direct them. Not only do negative emotions damage our psyche, but maybe they can adversely affect our physical well being as well. So when you are going down that path, take a second and think what could be the internal damage you are causing. Which, once again, leaves me baffled to why I experienced a stroke!

THE POWER OF MUSIC

I once read where there was a plant study done. They had plants growing in two separate locations with the exact physical conditions like light, soil, temperature and humidity with the only difference being in one room they had classical music playing nonstop. The results found were the plants in these conditions were healthier and grew larger.

Being a quasi singer, song writer, I often used music to convey the message I wanted to expound upon even though I got off to a most inconspicuous start. When I first started to learn how to play the guitar I was about thirteen. I had a surprise visit from a few friends who noticed my old beat up electric guitar.

They goaded me into playing a song, promising not to laugh. In their defence, I was quite raw and not very good and they couldn't help themselves, but to laugh and even teased me in front of others. When dealing with a situation like that we have options, improve and hone your skills, quit, or do what I did, for the next twenty years as I wrote literally hundreds of songs only sharing them with the four walls of the room I was in. The creation of a new jingle always brought me joy as I fantasized of performing on a stage.

This all took a sudden turn when a grade four class I was teaching found out about my secret passion, imploring me to perform for them using my own words against me. They used my own quotes such as, *"A HERO AND A COWARD ARE BOTH AFRAID EXCEPT A HERO TAKES ACTION WHILE THE COWARD DOES NOTHING." "COURAGE CAN ONLY HAPPEN IN THE PRESENCE OF FEAR."*

I was once described by someone as courageous when I played hockey. I tried to hit any opposing player despite their advantage in size. With all my chipped teeth, it proved to be more of an activity in stupidity than bravery being too small and slow to be that aggressive.

Courage was walking into that classroom with my guitar to perform for a group of nine year olds. I was extremely nervous, "My palms were sweaty, mom's spaghetti." I had a wonderful rapport with that class and doing this proved to be a most positive impactful experience.

They absolutely loved it as they were a bright class realizing I was the person responsible for writing the grades and comments in their report cards. But it does go to show where we can gain confidence and strength from the unlikeliest of sources, as I gained it from young children and just maybe you are the catalyst for someone else's rise.

I now sang at school assemblies and would visit random classes to perform to get messages across. This became an important staple in my life skills classes and when I would do bullying prevention presentations at other schools. I could sense when the audience was getting fatigued with my talking and would take out the guitar and crank out a tune to reenergize the room. It proved to be a successful strategy and I loved it! The coolest thing was I even did concerts with the band being former students and by far, I was the least talented, but the one who had the most fun living the dream as I would ask my wife how does it feel being a groupie to the lead singer of a rock band?

Needless to say I never got the response I hoped and anticipated. I wasn't a very good guitarist or possessed great musical chops but I could write catchy tunes and thoroughly enjoyed performing my latest creation to anyone I could convince into listening.

Thinking I was onto something, I came up with a masterful plan. With the support of the EGM Society, I wanted to go into schools and do bullying prevention presentations using music and humour to convey my message. I even propositioned schools that I would come visit a month early and work with students to have them sing and dance with me making it part talent show as well.

I was so enthralled with the possibilities, I retired prematurely so I could focus on my new career path. I finally got a gig at a junior high school in Beaumont. I visualized what a very successful presentation would be like and this was it! The student performers took it to another level and we all had a blast! I was certain this was my destiny and was totally thrilled with the potential of taking my show on the road. And exactly two weeks later, I had my stroke.

I believe in my heart I will be performing again and the best part is, people will be more tolerant with my lack of skill coming back from recuperating from a stroke. After working so hard in developing this program, I have no idea what the future holds but believe, "WHEN ONE DOOR CLOSES ANOTHER OPENS UP." I will just keep toiling along until another opportunity comes my way whatever it may be.

"Things don't happen to us, they happen for us!" ~ Tony Robbins

"Things happen for a reason!" ~ *Anonymous*

These concepts can be difficult to digest when you are going through a personal trauma. I do subscribe to the old adage, *"WHAT DOESN'T KILL YOU, MAKES YOU STRONGER!"* I have provided such a tiny snippet of examples of people overcoming overwhelming obstacles who have reached to what were perceived as unattainable heights. There are countless people all around us who have done likewise, but we are not aware of their story as they carry on with life in a dignified manner that we have no idea what they have battled through. And you very well may be one of them and probably are in your own special way.

We all have our demons that we need to overcome in order to become all that we were destined to be. It is probably something you wish you did not have to go through because who enjoys emotional turmoil? Other than grade eight girls. But, what happens when you go through rigorous physical training?

The next day, you will probably be experiencing body aches and pains which means the muscle cells in your body are in a state of rebuilding and growth allowing you to develop more strength. My first climb of the season would always result in pain compared with the ease and satisfaction I felt with my last climbs of the year.

I engage in reading material which is often considered controversial, but I believe you need to do what you need to do in order to move ahead without negatively impacting others. I believe there is a God and we go through a process in the evolution of the soul which can only occur in the face of adversity. I need to embrace this concept because it gives me the strength to push forward and establish a healthier and more productive point of view.

By accepting my stroke is a challenge and if I deal with it in a positive and possibly inspirational manner, I may become a better and stronger person has helped take some of the sting out from my current predicament. Trying to make sense of why bad things happen to good people can be an exercise in futility.

By believing things happen for a reason has transformed my mindset to one where I believe I have been granted an opportunity to prove to all my former students those words I spoke were true and I have the chance to demonstrate it and by doing so, I can further impact them and others in a positive way which I have always so enjoyed. Mark Twain once said, *"THE TWO MOST IMPORTANT DATES IN A PERSON'S LIFE IS THEIR DATE OF BIRTH AND THE DAY THEY FIND THEIR PURPOSE IN LIFE."*

Well, well, well, maybe I have discovered my purpose because without experiencing a stroke, I would not have written the two books. And this process has served to buoy my spirits and hopefully others.

RANDOM MUSINGS

These are random thoughts I would throw out there having no idea where they came from. In talking about fear especially with the fear of dying. Being a big believer there is life after death, I would state, *"I DON'T FEAR DEATH, BUT I AM MORE AFRAID OF NOT LIVING THIS LIFE TO ITS FULL POTENTIAL."* Meaning regardless of the predicament, we have to keep soldiering on and making the best of this life.

Another random thought, "WHAT IF IN HEAVEN, WE EXPERIENCE HOW WE MADE OTHERS FEEL?" This really struck a chord with me which I tried to embrace especially in the second half of my career.

When talking about someone who is not present; imagine they are standing right beside you. Would this affect your choice of words and tone of voice? The one thing that irritates everybody is someone who talks smack behind your back! Don't be known as that person and don't confuse honesty with mean-spiritedness. This simple practice will enhance your self-esteem because it demonstrates character.

Try to avoid putting yourself in a situation where you have to say, "I was only joking!" when someone is offended.

FATHER'S DAY 2020

In the midst of the chaos with the global pandemic and the Black Lives Matter Movement; for Father's Day, my daughter,

Megan and I with Dan and his father, decide to scramble to the top of Moro's Peak to celebrate the occasion. With my new situation, this proves to be a true slobber knocker as my balance is put to the test. I am very comfortable in professing my love and admiration for Mother Nature, but things got a little awkward and uncomfortable as I hugged her too many times for my liking. Good thing there was not a three knockdown rule!

When we make it to the summit, I am overcome with emotion because it is difficult for me to fathom I actually made it, but realize I still have to descend the mountain. There is a cross at the top. As once told to me, it was erected by a man named Moro who so loved this hike as the views are truly spectacular. He put the cross up to leave as his legacy when diagnosed with a terminal illness in honouring this heavenly slice of heaven on Earth! Which begs the question, "What will be your legacy and what will you be remembered for?"

Dan suggests I should do a video blog. He films me sitting by the cross and overlooking the majestic valley. I give a shout out to first responders and try to be a voice of inspiration and encouragement to those who may be struggling during these turbulent times. I post it on Facebook and the responses are so overwhelmingly positive from such a wide assortment of people. To me it reinforces how the majority of people have good intentions and are trying to get by the best they can despite being overshadowed by the negative actions of the few which seems to garner so much attention and dominate the news.

What a great day that will be treasured for years to come! And it was hard earned as with most special moments in life!

LIFE'S LITTLE INSTRUCTIONS PART TWO

- GIVE PEOPLE MORE THAN THEY EXPECT AND DO IT CHEERFULLY.
- DON'T JUDGE PEOPLE BY THEIR RELATIVES. With all due respect to my brother and sister. This one hits too close to home.
- CALL YOUR MOM because there will come a day when you can't and you will miss it dearly.
- SMILE WHEN ANSWERING THE PHONE especially when it is a friend who is taking the time to reach out to you.
- A LOVING ATMOSPHERE IN YOUR HOME IS SO IMPORTANT. DO ALL YOU CAN TO CREATE A TRANQUIL, HARMONIOUS HOME.
- READ, READ FOR FUN, READ FOR KNOWLEDGE. It is a well documented fact how successful people are avid readers. Coming from the man who is trying to sell books and striving to become a best selling author.
- BE GENTLE WITH THE EARTH
- REMEMBER THAT NOT GETTING WHAT YOU WANT IS SOMETIMES A STROKE(UGH, that word again) OF LUCK.
- LIVE EACH DAY LIKE IT'S YOUR LAST because one day it will.

BUCKET LIST

I would have students make a list of things they would like to accomplish long before the release of the movie starring Jack Nicholson and Morgan Freeman. In A 'Chicken Soup for the Soul' book, I read an article about a man who I believe was John Goddard, who wrote such a list when he was fifteen and had over a hundred items and he crossed off almost all of them. I would read some of his list to the class and of course, highlighting the ones I had managed to do myself.

Apparently, making a reference to kicking the bucket is appropriate, but I offended a parent with my title heading of 25 things I want to do before I croak. I suspect it had to do more with my light hearted attitude towards death and dying rather than their affinity towards amphibians.

Sheesh, the perils of trying to be humorous when you obviously are not as by now you can attest to, but I do try. I would get my students to write down twenty-five future accomplishments and I would read them to pick up ideas for my own agenda, as they were quite often creative except for wanting to party with me in New Orleans or having me on the reality show, Survivor.

The great thing about having such a list is when you start scratching off activities you have successfully fulfilled, you realize how anything is possible and you get excited about moving onto the next one, making life that much more interesting and satisfying! Quote time!

> *"If you fail to plan, plan to fail."* ~ Anonymous

It always brought me great joy when running into a former student and hearing them joyously proclaim a number of how many things they did off their bucket list. The cool thing is, as we continue to grow and develop, so does our list evolve opening the door to new and unthinkable possibilities.

Now to be honest, I have never actually put down my future to-do list on paper and neither do you if you do not want to, but I would highly recommend you spend time thinking about it and creating a list in your mind. I have had to completely revise mine, but it does not make the new list any less exciting.

Your list doesn't have to impress anyone else but yourself. For example, Souchie's dream round of golf would be at Pebble Beach and for me it would be at Deer Meadows. Because it's a short nine hole course close to my house and this would mean I can swing a golf club with two hands again. When that happens, I will reconfigure my list and join Souchie at Pebble Beach.

OPRAH WINFREY

It would seem inane to write a book such as this without referencing quotes from the queen of inspiration and compassion. Her forays into this field have made her a legend and deservedly so. She has delivered some marvelous insights and this first one is one I use quite regularly.

> *"Doing the best in this moment puts you in the best place for the next moment."* ~ Oprah Winfrey

I would think about this during leg day at the gym and not wanting to do squats which I know will make hiking in the mountains easier or when my rehabilitative exercises became tedious. It is like doing extra training for work which can lead to opportunities in the future especially when the courses seem boring. I marvel at the tenacity and discipline of my daughter; while holding down a full-time job, is taking courses online to receive her masters degree in counselling so she can gain better control of her future in the workforce to ensure career satisfaction.

> *"Challenges are gifts that force us to search for a new center of gravity. Don't fight them. Just find a new way to stand."* ~ Oprah Winfrey

Which is something I have literally had to learn how to do.

> *"The more you praise and celebrate your life, the more there is in life to celebrate."* ~ Oprah Winfrey

Sounds like reflecting on daily delights!

> *"Breathe. Let go. And remind yourself that this very moment is the only one you know you have for sure."* ~ Oprah Winfrey

> "So go ahead, fall down. The world looks different from the ground" ~ Oprah Winfrey

TECHNOPHOBIC MAN

Technophobic Man is a biographical song I wrote which expressed my dislike and frustrations with the world's preoccupation with technology. It is a quirky little tune which I thought was terrible even though it became the most requested song from children. Coming from a man who could not have written his books without the use of a computer and who is someone who has a website, is on Facebook to promote his material and sells his books on Amazon. And here is my main rationale:

> "I fear the day that technology will surpass our human interaction. The world will have a generation of idiots." ~ Albert Einstein

The point being, someone may reach out to you with kind words and make you smile with a humorous message, but that does not supplant a hug or laughter shared together with friends. [I know we are in the midst of a global pandemic, but do you remember the warm fuzzy feeling you got from a heartfelt hug?]

Is it a coincidence with the advent of technology in our daily lives that mental health issues are on the rise and the struggle for young teens to find their way is more difficult than ever before?

> "Love and compassion are necessities, not luxuries. Without them humanity cannot survive. We can live without religion and meditation, but we cannot survive without human affection." ~ Dalai Lama

I felt there were administrators who thought a lot of the programming I offered were just fluff activities merely to amuse children, not fully understanding the components of human interaction and character development which were the integral focal points.

In my eyes, it feels like technology has diminished the capacity for some of our youth to embrace compassion for others ergo the rise in cyberbullying. People can write things about others with no regard to how it may affect another. Even I was not immune to it, but it was a good thing I was an adult and could put it into perspective and deal with it.

I saw a video where students were using plastic cups as percussion instruments for a song. I approached our wonderful music teacher, Josephine, and we came up with a plan to get the whole school involved to enhance the school spirit. [Could someone please explain to me why every school does not have a specialist music teacher and music program?]

I have had the privilege of working with two amazing ladies, Josephine and Susan, who captured hearts and positively impacted so many struggling children). Josephine created percussive arrangements to an original song I had written which she worked with students during her music classes.

The grade nines came up with the idea to write short positive messages for the other students to read when they turned their cups over. It was an exciting day when we pulled it off and the school was buzzing for many days after. We put the video on YouTube and it garnered a wide array of responses.

Now I didn't realize that you were only supposed to do this with a specific song known as "The Cup Song." So when some people logged onto the video and saw something else, they felt it was well within their right to voice their displeasure. Someone commented this was the worst song they had ever heard! Now this rattled me because I held such a low opinion of Technophobic Man, I thought to myself. "Could I actually be responsible for two of the worst songs ever written?"

Someone else described it as a piece of fecal matter! I was doing a presentation at a junior high school with the grade sevens sitting in the front row. When I shared this story, I could see them asking each other what was fecal matter? Being the professional I was and never missing the opportunity for a teachable moment, I calmly told them to ask their science teacher. It just makes you wonder what is the sense and purpose for spewing such negativity? And yet, this happens countless times every day around the world to vulnerable souls. It simply has to stop!

Another pet peeve with technology is it can really dampen the spirits of those who are struggling. I was told I needed to be on Facebook to promote my first book. I understand the whole point of Facebook is to share things with friends. It can be hard for people who are struggling to see how awesome everyone else is seemingly doing when you are stuck in a rut or in a low place.

I would see pictures and read stories of people on exciting adventures when the highlight of my day was getting dressed all by myself. This is just another classic example of the perils and pitfalls when you spend your time comparing your life with others rather than focusing on what you need to do to enhance your own. If you are not in your happy place, I would offer one should devote their thoughts and energies towards creating a special memory with someone else rather than spending time on a device. Which leads me to my favourite cartoon caption which has two guys in a conversation at someone's funeral with row upon row of empty chairs. One says to the other, "I thought for sure there would be more people here. He had over 800 friends on Facebook."

Okay, enough already! All I am trying to say is we have to be responsible and sensitive towards others with our use of technology and we cannot let its usage upstage or supplant our need for human interaction. And now for my take on virtual schooling! You can relax because I am going to stand by my words enough already.

YOU GET WHAT YOU GIVE

This was my go to phrase especially when my grade fours were impressed with my popularity in the school. I explained to them it was not always the case until I made the giving of high fives a regular part of my day. It proved how initiating good will towards others is such an easy gesture with huge rewards.

Interestingly enough this was Courtney's favourite quote and she totally embodied this. I was asked to do a virtual bullying prevention presentation for a school in another division. Our high school students had the day off and I asked a few of them to join me because I did not think I could pull this off on my own.

Without hesitation, Courtney willingly accepted my invitation to participate. We really didn't rehearse anything so I took a chance and threw a question out to her knowing the response would probably be quite positive. I asked her, "In your school experiences, how many times have you been bullied?" Her response was pure gold because here was her unscripted answer, "I have never been bullied. I get along with everybody. I try to treat everyone the same no matter how different they may be." And to all who know her, this is so believable and what a great message for the session!

And here is Phillip's astute take, "I have seen from the classroom to Florida and back now, how important these five words are. Every day the world throws curveball after curveball, but if we give it our all and leave everything on the table. The reward is pretty amazing in the end!"

BARRY'S JOURNEY

We were driving down the highway when I turned to my wife and said, "You know I've been lucky."

She turned to me and sternly declared, "You can't call yourself lucky. People who have cancer can't say, "I'm lucky."

Not the car ride, but the cancer journey, started some months earlier. It began with Dr. Melissa Johnson saying, "I'm sorry to tell you, Barry, but we found cancer." She was sorry. Kathy burst into tears. So to consider myself lucky some months later does seem to fly in the face of reality.

There's this old saying, "What doesn't kill you makes you stronger." It always made sense to me in that life's challenges prepare us to face the next challenge, making us mentally (and perhaps physically) stronger. Facing cancer though gave rise to a different strength: gratitude.

My gratitude was profound. Family, friends, and the people who work in the delivery of cancer treatment all left me grateful. I could say thank you, but those two simple words were almost trivial to the deep sense of appreciation that I felt because of the care I was given.

In many ways, the strength of gratitude left me a more aware person. We can easily take those around us for granted. Gratitude led to appreciation of family and friends whose outreach and engagement was humbling. Gratitude led to appreciation for the medical staff and scientists whose work led to me getting a voice mail from Dr. Joseph, "Barry, I'm just calling to tell you that your CT scan showed you are cancer free."

As often happened on this journey, gratitude gave rise to momentary tears of joy. They were tears that come from the strength of gratitude. Yes, I'm damn lucky.

by Barry Gibson

"BE SOMEBODY!"- Josephine Ivancic

These were the words my mom would regularly say to her children and grandchildren. When my mother passed, my daughter got a tattoo with those words on her left forearm as a tribute to her grandmother. They loved each other dearly and as a father, it was a true blessing and joy watching this relationship blossom and flourish as there was no shortage of laughter when the two of them were together. Case in point, when my mom visited one year and I dropped the two off so they could try bowling; my mother was in her late seventies and this would be her first foray into this sport. On the drive home, I asked my daughter for my change, so she returned all my money. I was about to chastise my mother for paying when she declared she had not paid either. A quick U-turn and I was back at the bowling alley with both felons too embarrassed to join me. Talking with the attendant, I explained to her how I was totally fine if she wanted to contact the local police and that Thelma and Louise were in the backseat of my vehicle.

To my mother being somebody never entailed being rich, famous or possessing some type of social status, but it meant you were a contributing member in society and you were making a difference no matter how trivial or small it may seem. She modelled this attitude when she worked at a seniors home.

She would serve the residents their meals and would go out of her way to ensure they had the little extras that brought them delight as subtle as an extra cream for their coffee or a specific type of marmalade for their toast. These simple acts endeared her to them which goes to prove how it's in the small details that can make a huge difference! Needless to say when my mother retired, the residents were quite upset. Every little action we take to improve another person's day is truly being a somebody. Quite simply, BEING A SOMEBODY MEANS BEING A SOMEBODY FOR SOMEBODY! And this something we all can do.

MARY, BASKETBALL AND A LITTLE STREET CRED

It's not easy to put into words how much a mentor means to someone. As I sit here and try to write a submission for this book, I find myself sifting through an abundance of memories trying to choose which anecdote would fit best. Then, I catch myself thinking of how incredibly lucky and grateful I am to have a person in my life who has impacted me so deeply in so many different ways that I can't choose just one moment to talk about. That being said, I will do my best to convey what Mr. I means to me, and what the name "Mr. I" means to so many.

A quotation from a personal essay that I wrote in grade 12- December, 2007:

"The coach's name was Mr. Ivancic, but everyone calls him Mr. I. He is the teacher that every child in the school loves because he has a type of charisma that draws

people to his kindness. He also cares about every individual for exactly who they are."

I was in grade six when I moved to École Meridian Heights School; this was long after Mr. I had established his name and his obvious knack for impacting students positively at this school. He was known and loved by so many, and his energy vibrated through the hallways. We first established a true connection when he accepted me onto the junior girls basketball team in grade seven, although I had absolutely no experience. At this time, I was excited to have the opportunity to make some friends and be a part of a team, little did I know that it would shape me as a person, and quite literally, influence the course of my life.

Basketball has certainly altered my life in terms of my time playing and coaching, but I am specifically referring to the life lessons that I have been taught through this beautiful sport- all starting in the hands of Mr. I. He somehow made being a part of the Meridian Heights Magic team be just that- magical. "Respect every single person who walks into that gym," he would say. Teaching us not only that we must respect our opponent, but to be gracious to officials and to be kind in the face of adversity. This lesson in humility has extended far into my playing and coaching career. It is something that I consistently try to instill in the athletes that I am fortunate enough to work with. It is also something that I remind myself of as a teacher and a person. Acknowledging that all people deserve respect sounds like a simple notion, but it is not always what we see in our world. I have strived to take this life lesson and spread it to the youth that I work with and to reflect it in my actions. I truly believe that all people deserve respect and kindness, and what a great way to teach that to children through sport.

My personal favourite line that he would recite in our basketball practices and games was "The easiest thing to do in life is quit." Applying this to sport was one thing, but applying this to endless moments in my life has helped me push through situations that I never thought I would overcome, or in some cases, survive. I, like so many, battled mental illness silently for years. It wasn't until another brave teacher in my school division spoke out publicly about her experiences that I decided to open up to the public about my struggles with episodic depression. I have been living with episodic depression for over a decade, and many of these lessons- seemingly small basketball lessons- have helped pull me from some very dark moments and brought me back to more manageable spaces. I continue to fight this battle with mental illness not because it is easy, or because I always have the fight in me, but rather because I know I can. "The easiest thing to do in life is quit" rings true in my dark moments. Countless times I have wanted to quit and submit to the all-consuming hopelessness. However, many people, including Mr. I, have shown me that I am so much more capable than I could ever give myself credit for. Not quitting in these times feels impossible to me, but I always know I can keep going. Through basketball, Mr. I taught me resilience and perseverance; two qualities that I rely on daily to embrace my struggles and all that comes with them in order to live the fulfilling life that I have.

I have learned countless lessons through sport throughout my life. I could go on and on about specific moments, quotations, and memories that have influenced me so deeply. I feel an immense amount of gratitude for being introduced to basketball in grade seven. It has been 18 years since a very special person introduced me to it; I will hold Mr.I near and dear to my heart forever. It's not easy to put into words how much a mentor means to someone. As I sit here and try to write a submission for this book, I find myself sifting through an abundance of memories trying to choose which anecdote would fit best. Then, I catch myself thinking of how incredibly lucky and grateful I am to have a person in my life who has impacted me so deeply in so many different ways that I can't choose just one moment to talk about. That being said, I will do my best to convey what Mr. I means to me, and what the name "Mr. I" means to so many.

> *"Life is not a spectator sport!"* ~ Mark Ivancic

Unfortunately, when people are not doing well, the tendency is to pull back and finding the energy to do the simplest of tasks may present as a daunting challenge. Any action, however, can serve to be the tonic for whatever ails the hurting heart. Here are some random quotes that hopefully illuminate the need to take action.

> *"You don't stop playing because you get old. You get old because you stop playing."* - Anonymous

> *"If you don't have confidence in yourself, get off your rear end and do anything that will make you feel better about yourself."* - Anonymous

> *"People are always blaming their circumstances for what they are. I don't believe in circumstances. The people who make it in this world are the people who get up and look for the circumstances they want and if they can't find them, they make them."* ~ Anonymous

> *"Interesting people do interesting things. They take an interest in the world around them and constantly try to improve their lives and the situations around them."* ~ Anonymous

I know for a fact there are a lot of young people who are struggling, which is why I felt compelled in offering different types of programming which focused on meaningful social interaction and hopefully some character development. It is inane and archaic thinking to believe this magically occurs for all students within the four walls of a classroom.

We need to be creative in creating scenarios to foster growth as we do in our regular day to day lives. The reason why trying new things is imperative despite our age brings me back to my wife's grandmother who possessed an endearing sparkle and zest for life. After chatting with her, you walked away admiring her spunk and wanting to do something. She was in her eighties and tried learning how to ride a bicycle again. She asked me if the reason she kept tilting to one side was due to an issue with her balance. I just smiled and thought to myself, "No, it's because you are eighty-three years old!" In essence, trying new things adds spark to our lives and keeps us young at heart. The bottom line is that it makes the expedition more fun and enjoyable!

> *"Just do it!"* ~ *Nike slogan*

Trust me, I had to be ultra careful in my wording when discussing this slogan with junior high students.

> *"Inaction breeds doubt and fear. Action breeds confidence and courage. If you want to conquer fear, do not sit at home and think about it. Go out and get busy."* ~ *Dale Carnegie*

"I GET BY WITH A LITTLE HELP FROM MY FRIENDS" Is a title from a Beatles song which Joe Cocker re-released as a single and is arguably better than the original version even when the Beatles were at their peak which goes to prove even when things are going well, there can always be a tweak to make things better!

The inspiration for this came from a conversation I had while texting my sister one sunny afternoon. And some of you may be thinking. "Texting??? And there you go Mr. Technophobic Man!

"I was telling her how grateful I was for her enthusiasm towards this project. When you embark on something like this, you cannot help but question yourself whether it is as good as you think it is. But Natalie without ever having read a page was always so excited for me and kept encouraging me to carry on. When I told her how thankful I was, she replied, "We can all use a little support from time to time." Which triggered the name for this entry. And here is the trick question of the day.

IS IT BETTER TO BE THE GIVER OR THE RECIPIENT?

You can't depend on being a recipient in your time of need but you can strive towards being a consistent giver in someone else's time of need which can be as simple as an encouraging or positive word.

The best part about this is that I wanted to address the importance of friendship in a different manner from my first book and lo and behold, I got the inspiration from my sister which goes to prove there is such a thing as cosmic humour.

> "Only in the dictionary does success come before work."
> ~ Anonymous

> "The harder you work, the easier things get." ~ Anonymous

It seems in our world of instant gratification putting in extra effort or time is construed as a negative when that is truly the recipe for success. I have had numerous people tell me if they wanted to get to the top of a mountain, they would either rent a helicopter or open a can of Coors Light. Somehow I do not think that can compare with the feeling of accomplishment of actually doing it as I witnessed first hand with my grade nines.

Sorry, I can't help myself. Time for another grade nine mountain hike story. We get to the top and I see a boy off to the side talking on his cell phone and he is in tears. I overhear him talking to his mother when he tells her, "Everyone kept telling me I couldn't make it to the top, but mom, I am standing on top of a mountain!"

His mother was a co-worker with a friend of mine who told me she overheard the same conversation at the other end of the phone line as the mom broke into tears. You can't make this stuff up!

This summer I embraced this challenge by getting to the top of four summits in the mountains. I would post pictures or a video after each accomplishment and I would receive all kinds of wonderful and positive responses. Some would hail me as an inspiration but honestly, I was just trying to impress myself. After each success, I expanded the possibilities of what I could attempt next. While there were others who I felt, thought I just sprinkled pixie dust on my butt enabling me to magically float to the top, not realizing how difficult it truly was for me with my left arm dangling at my side proving to be more of an impediment, as I could not use it as a support when navigating through certain spots which required balance. Or even the fact, it feels like I suffered nerve damage to my hip and I have to walk with a much slower and awkward stride making the excursion that much longer. But it did enable me to practice social distancing as I lagged behind. Just like people may not know how hard you are working to keep your head above water, but you keep taking steps forward so you can enjoy the fruits of your labour. We do it to impress ourselves and because it is what needs to be done.

Achieving goals often involve struggle and discomfort which explains why the personal victories taste that much sweeter. Have you ever noticed when upsets happen in sports, the primary reason explaining why this occurred was due to the opposition having outworked the perceived favourites? The thing is we all have latent talents and when we combine them with a superior work ethic, life becomes that much more interesting with more success and personal fulfillment.

We need to reframe the concept of work in our mind from an onerous task to an opportunity to better oneself. I am baffled with how many times in youth sport today, families treat practices like an inconvenience. The young athlete proclaiming they would much prefer just playing in games rather than hone their skills working in practice.

I read once where the elite players in a hockey game on average handle the puck for thirty seconds during an entire match, but in a practice an athlete could be working with it for an entire practice which usually leads to more success in games, translating to more enjoyment and fun which is the whole point of playing sports.

Of course, this radical shift in my attitude stemmed from my transition from player to coach, but we should all be constantly evolving to make our lives more successful. When faced with the challenge of physical labour, as mentioned before, it is an opportunity to work on personal fitness.

School is boring? Boo hoo hoo! You are being given an opportunity to pursue your dream job and the career of your choice. What an incredible luxury our society provides! I would share with my classes the difference of the university experience between med students and those in the education department and why doctors deserve every positive perk life bestows upon them. Simply put, they have EARNED IT!

If school is a struggle, I would suggest reading as much as possible and especially this book. My daughter once shared a valuable insight with me on how to do well in school. "It's really not that hard Dad, if you do these three things; pay attention in class [What? Even if the teacher is not wearing a clown suit and performing to ensure the lesson is not boring? Coming from the man who would readily do that.]

I flourished under the concept of being paid to be the class clown and would tell those students who tried to supplant me how I had earned the right to tell bad jokes in a classroom by receiving a university degree and if they wanted the opportunity, they should do likewise. Even if my father regularly reminded me, "Five years university and you still don't know nothing!" In his defense, I did give him due cause to make that statement when he deemed it fit.

Secondly, do your homework and study for tests. In other words, put the time and effort in to ensure success. Which is pretty much the formula in achieving any goal in life you desire because it is not work, it's OPPORTUNITY!

Speaking of my university experience, I cannot help but think I underachieved in three areas; as a student, as an athlete and as a person. Don't get me wrong, I met so many outstanding people who I shared wonderful memories with that I still relive to this day and probably to the surprise of more than a few, I did graduate. The thing is, there will be times when we feel like we have underachieved, learn from the experience, move on and don't let it happen again.

There have been more than a few occasions in my life where I have overachieved and I am sure it had a lot to do with my commitment to doing what it took in making the endeavour a success.

Recognizing and owning it when you have underachieved is an important first step to creating positive change and ensuring it doesn't happen again. And for those of you who feel I have underachieved with this book, I would suggest you keep that to yourself or else I will feel compelled to try again.

> *"Opportunity is missed by most people because it is dressed in overalls and looks like work."* ~ Henry Ford

> *"The heights by great men reached and kept were not attained by sudden flight but they, while their companions slept, were toiling upward in the night."* ~ Henry Wadsworth Longfellow

THE CHILDREN OF ECUADOR

The Children of Ecuador was a program started by my good friend, Dave Oldham. He was doing some wayward travelling when he stumbled upon a quaint village in this impoverished country and did volunteer work at a local school. The experience left an indelible impression on him prompting him to start a leadership program at the high school he was working at with the focus being on helping the children in that community gain an education. As you shall soon read, he took groups with him to experience the power of doing service in a foreign country which is something I would highly recommend everyone put on their bucket list.

My daughter took the class and was given the chance to visit this majestic country. She wanted to go, but had one stipulation and that was I accompany her. I was thrilled because not too many dads get such an offer from their seventeen year old daughter to experience a thrill of a lifetime together. And for the record, my role was not as a wingman but more chaperone and bouncer if need be. It is now time to share Dave's wonderful description of this program and experience.

The Hill

As the bus rambled up the road from the coast of Ecuador, anticipation was building as our group of 47 volunteers took in the ever changing landscape of a foreign country. 30 minutes passed as the rolling hills of the farmland became ever present in stark contrast to the coast. "Aqui aqui" (here here) I said as the bus came to a halt near a narrow dirt road that didn't seem to have a destination.

Three years ago I had first walked this road to Escuela Una de Octobre with my sister. I'd made that walk 100 times during our three month volunteer teaching experience. Now I'm back with a large group of eager faces to show them what captured my heart. Our group ranged from 16 to 60, from experienced traveller to first time passport holder. Everyone had their own reasons for joining our goodwill mission to Ecuador, but we were united in the desire to make a positive change in the lives of others. We also spoke often about perspective and gratitude. Coming from Canada, everyone in our group has greater wealth by several orders of magnitude than anyone we will encounter on our 10 day goodwill trip. How can we leave this place better for our being here? What will we take from this experience that will leave a lasting impact in our own lives?

It's a beautiful walk up to the school in the hills, and often times really hot! On this day, 35 degrees and humid with not a cloud in the sky. Bamboo huts dot the vast fields where our Ecuadorian friends work tirelessly to provide for their families. For years, their children would never attend school as it was deemed a distraction from the work. Now, with the lions share of the effort being done before I had ever arrived, a small school, cookhouse, playground, bathroom, and of course soccer field is now the central gathering space for the entire community of La Florida.

As we crest the final ascent to the school, the sounds of children playing are familiar to all of us. Our group is at first shy and tentative. "What do we say?" "If only I had learned more Spanish!" I let this awkwardness simmer for a few minutes knowing the contrast over the next few hours will change some lives. In the hours and days following this first interaction, we connected, laughed, and even shed a few tears. We did this through painting classroom walls and building a fence. We donated supplies and provided backpacks. We taught the children how to skip rope and blow bubbles. We lost the first ever Canada vs. Ecuador soccer game in the history of La Florida. We started in small groups of Canadians only, taking on a project and thinking the goal was to finish the work. Soon enough, the inquisitive local children would grab a paintbrush or hammer a nail. Within an hour we would all be working together, which was really the goal. Laughter and broken Spanish filled the air as friendships quickly formed. By the time we said our farewells, the uneasy shyness had given way to an Ecuadorian dance party, hugs, tears, and promises of return (and we did!).

My favourite part of any Ecuador trip is the walk back down the hill. The conversations are different. They are personal, and come from the heart. Something about the experience at the school and with those kids opens people up. The walk up the hill is filled with anticipation. The walk back down is filled with gratitude and a new perspective. I've taught and coached kids for 25 years in my hometown of Spruce Grove, but some of the most meaningful conversations I've had happened walking down that hill in Ecuador. The teenagers on the trip talk about how this experience will inspire them to travel more, volunteer with joy, and find their own way to change the world. The adults talk about how grateful they are to do this with their kids. Those that don't have kids on the trip talk about wishing they did this earlier in life and no longer waiting on the sidelines. We all talk about wanting to do more for others and the impact that giving to others has on you. It truly is the ultimate win-win.

So, go find your hill. You likely won't even know you've found it until you are on the way down.

By Dave Oldham

Dave is so right when he describes it as the ultimate win-win which are scenarios we should try to create as often as possible in our daily lives. A true win-win is when everybody walks away from the experience in a positive state and there are benefits for all. The Ecuadorian students gained from the material objects we provided while ours profited in profound ways too complicated to measure but of unlimited importance for the soul. It was this realization where I had the epiphany which spurred the creation of the Unity in the Community program.

As Dave and I were floating down the Amazon River in a canoe after having visited a wildlife refuge in the Amazon jungle, I had another idea where I could do a fundraiser by hosting a charity concert with all the funds going to support this tremendous cause. A Win-win scenario? For me? Yes because I had a reason to live out my rock and roll fantasy. For this organization? Yes as well, as we did this several times and managed to raise a nice chunk of money for this worthy cause. For the audiences? Maybe not so much as they had to pay their hard earned money to hear me attempt emulating a rock star. I never felt bad about that though because I always surrounded myself with talented musicians, fantastic vocalists, our school choirs who were quite often provincial champions and a lot of positive energy making it a true celebration.

MANAGING THE VOICES IN YOUR HEAD

Whenever we embark on a life challenge or opportunity, we often have different voices in our head directing us on how we should handle the situation. There is the voice of reason, fear, self doubt, discouragement, cockiness, discretion, confidence, self assuredness and positive belief to name a few. Once again our job is to sift through it and listen to the voice which will help us attain our goal.

I had a job interview for a supervisory position with an organization I was working for. To prepare for it, I read an article which said you have to convince them you are the best person for the job and to portray an image of great confidence with your abilities. I listened to the voice of vanity and came across as an arrogant and pompous fool and did not get the position. For those of you who are sarcastically saying to yourself, "Shocking!" I want you to hold that thought as you read the rest of this entry.

Apparently, I was the leading candidate for the position until I started talking. Not the first or last time, this would be the scenario that would play out for me. As a basketball coach, I would tell my teams you have to start making new mistakes instead of the same ones over and over again so I can do a better job of coaching. Point being, we have to quit making the same mistakes in our lives.

No word of a lie, this is what transpired in my very next job interview. It was for a teaching position in a Catholic school division and I had a person to person meeting with the superintendent. Things were seemingly humming along quite smoothly when I got asked THE question. It was in the late eighties and he asked me what I thought of the pope's stance on AIDS, "Bamma ramma slamma jamma. I got this!" I thought to myself when I listened to the voice of bad humour and poor discretion. So I replied, "Sir, I imagine the pope believes every teacher should have a teacher's aid in their classroom." Apparently once again, wrong answer.

More often than not, we sabotage ourselves when we listen to the critics who seem to occupy the biggest space in our heads. Now this quote from Theodore Roosevelt is a classic and one we all need to take to heart.

> *"It is not the critic who counts; not the man who points out how the strong man stumbles, or where the doer of deeds could have done them better. The credit belongs to the man who is actually in the arena, whose face is marred by dust and sweat and blood; who strives valiantly; who errs, who comes short again and again... who at the best knows in the end the triumph of high achievement, and who at the worst, if he fails, at least fails while daring greatly!"* ~ Theodore Roosevelt

Which I have to constantly remind myself while trying to piece this book together. We need to listen to our best supporter voice and just go for it! We need to be our biggest cheerleaders without listening to the voice of vanity which should be the voice of self confidence that reinforces the belief I can and will do this because I am willing to learn and grow to do whatever it takes to be successful in this endeavour that is meaningful to me.

And get rid of the "yeah but" voice. Have you ever noticed with some people when you try to encourage them to try something to help them to move forward? Their first response is "Yeah but." I have firsthand experience with this because I went through this for a period of time and it did me no favours. To quote my great friend Curse, "Don't ask why? But why not?" In other words no excuses and just go for it. And so on the topic of excuses, I offer these.

> *"We have more ability than willpower, and it is often an excuse to ourselves that we imagine that things are impossible."* ~ Francois de la Rochefoucauld

> *"Ninety-nine per cent of the failures come from people who have the habit of making excuses."* ~ George Washington Carver

> *"He that is good for making excuses is seldom good for anything else."* ~ Benjamin Franklin

All this suggests is rather than focus on why you can't, get on your horse and make it happen. Like you have a horse? And if you do, you would tell me, great idea.

COMPASSION AND THE DALAI LAMA

As we work our way through the social dilemmas the world is currently undergoing, it would seem there has never been a greater call for compassion than now. And who better to quote than the Dalai Lama who has written many books with the central themes being compassion, the power of love and empathy?

Teacher time as for today's lesson we are going to study the definitions of compassion and empathy and you will be graded on your ability to put these terms into practice.

> *Compassion is not the same as empathy or altruism, though the concepts are related. While empathy refers more generally to our ability to take the perspective of and feel the emotions of another person, compassion is when those feelings and thoughts include the desire to help.*
>
> *While cynics may dismiss compassion as touchy-feely or irrational, scientists have started to map the biological basis of compassion suggesting its deep, evolutionary purpose. This research has shown that when we feel compassion, our heart rate slows down, we secrete the "bonding hormone" Oxytocin, and regions of the brain linked to empathy, caregiving and feelings of pleasure light up, which often results in our wanting to approach and care for other people. greatergood.berkley.education*

This all harkens to walking a mile in another person's shoes and imagining how another person would feel based on your actions or words towards them. It is an inevitable fact children will get under your skin and frustrate you. So before reacting I would try to remind myself, "How would I like to be ten years old and have a moustached version of Uncle Fester growling at me?"

And now for some treats courtesy of the Dalai Lama who I want to believe would approve of this message and the concept of this book. If he doesn't, I will just refer to the Theodore Roosevelt quote.

> *"The goal is not to be better than the other man, but your previous self." ~ Dalai Lama*

> *"Be kind whenever possible. It is always possible." ~ Dalai Lama*

> *"People take different roads seeking fulfillment and happiness. Just because they are not on your road does not mean they have gotten lost." ~ Dalai Lama*

> *"Happiness is not something ready made. It comes from your own actions." ~ Dalai Lama*

> *"Just one small positive thought in the morning can change your whole day." ~ Dalai Lama*

And this is why I encourage you the reader, to start your day by reading a page or two of this book to get the ball rolling. If you are a teacher, my hope is by sharing a passage with your class that it creates a positive start to the school day and that it serves to be the 1% positive difference in at least one student's day. Now that would be cause for a bamma ramma slamma jamma!

ROLE MODELS

We all need models in our lives and no, I am not talking about Christine Brinkley (Once again I reveal my age) or Melania Trump. Now the mere mention of Melania's name may have triggered a negative reaction by some because we are a reflection of who our friends are and the people who we associate with. And if this is the case, based on who my friends are, it makes me an incredibly cool dude! Without ever having interacted with Melania, it would seem unfortunate and unfair to pass a negative judgement upon her. Yet due to social media, this is seemingly the norm in today's world.

I am talking about role models who can be described as people who someone admires and whose behaviour they try to copy. Whether we like it or not, we are all role models as people, especially children, are watching what we do all the time.

> *"A role model in the flesh provides more than inspiration; his or her very existence is confirmation of possibilities one may have every reason to doubt, saying, yes, someone like me can do this."* ~ Sonia Sotomayer

> *"Each person must live their life as a role model for others."* ~ Rosa Parks

> *"My biggest thing about being a role model is whatever I am preaching, I am practicing."* ~ Gigi Hadid

This really resonated with me because I always believed if you talk it, you better be able to walk it. Which is why hockey players who talk tough when they are separated by officials garner little respect. I know teaching this class made me into a better person. Once again, I suggest if you are an educator, share these with your students and see if it makes you into a better teacher.

I would tell students to emulate the qualities of people who they admire, combine it with their own uniqueness and you create a special brand of awesomeness that the world has never witnessed before. It is totally okay to have people think, "You aren't like the other ones are you?"

Michael Jordan has a quote which states how you shouldn't have negative role models, but I disagree. I believe we should have them and should feel compelled to do the opposite of what they do because we may in fact have similar qualities we need to work on.

When I first started coaching basketball, it seemed like we were getting beat by thirty to forty points per game which is quite substantial for junior high basketball. Part of the problem I suspect was my transition from hockey player to basketball coach. Seeing there was no penalty box for a transgression, I would tell my players, "You get five fouls, so use them."

Not understanding, when your team picks up five team fouls, the opposition gets to shoot free throws for every ensuing infraction. No, ignorance is not bliss! I was not pleased in the least with these results, so I knew I had to change. I started attending coaching clinics to learn what it took to be successful and going to university basketball games always sitting behind the opposing team's bench.

I analyzed the different coaching styles and tried to imitate what the good coaches (who I wanted to be like) did. It was quite humbling to see a coach act in a way I thought was unbecoming and realize it was something I would do and realized this was not acceptable and that I had to change. This significantly altered my performance and the team success we enjoyed. I believe this is a practice we can use in our daily lives to improve our day to day performance remembering being a role model does not mean we have to be infallible but that we are in a constant process of self improvement and that is the lesson our children need to see.

On a side note, while talking about coaching basketball, I would recommend if you are a coach, you should get a dog. After every loss, I would take my dog for a walk to clear my head and try to sort out what I needed to address so we could move forward. I always returned home with a new strategy in place, enjoyed some fresh air and exercise, felt optimistic with the changes I believed would help the team to be more successful and I had a happy dog which is always a good thing. Now that I think about it, this would be a good practice for whatever is troubling your soul.

> "Life is not measured by time. It is measured by moments." ~ Armin Houman

The Buddy Mobile

When I reflect on this quote, I'm taken back to the 1990's. A decade that many would consider a much simpler time in the world. When you're a kid, life is full of excitement and wonder. You live for the thrills and over time you realize that it's the little moments in life that stay with you. This is my story about the Buddy Mobile. Naturally, as kids, we weren't permitted to ride in the front seat of most vehicles... what a bummer! However, this wasn't the case when it came to the old blue Corolla . The oldest, ugliest vehicle I'd ever seen would end up being one of my favourite childhood memories. You see, because this car was so old, there were no airbags in the front seat. This presented me with a very special opportunity to ride shotgun anytime my parents took old blue out for a ride. But, it wasn't just the fact that I got to ride in the front seat that made this car so special... it was much more than that. When I think of this old, blue Corolla, I picture myself as a young kid blasting songs on the radio and singing with full force beside my dad. We would cruise around the neighbourhood in this car with the windows down, rocking out as if nothing else in the world mattered... and in those moments, nothing else did matter. There was this one song in particular that would often play on the radio... a popular song by Train called Drops of Jupiter (which my dad would later argue was called Tears of Jupiter!). This song would come on and we would instantly crank up the radio and belt out the lyrics. In particular, we would be singing the song and when the following lyric would come up "Your best friend always sticking up for you, Even when I know you're wrong?" we would look at each other, pointing fingers at one another and burst out in laughter.

The point is, this old blue Corolla and the memories we shared together meant more than any fancy, new model vehicle, ever could. I remember this story so vividly because it was a moment that brought me such joy as a child. You can have as much time in the world with somebody but if you don't make the most of it, you'll have nothing to look back on. These memories will stay with me forever and I'll always be grateful for old blue. It's up to you to decide whether or not you want to explore the potential of the little moments in your life and turn them into longstanding memories. Sometimes I still hear a familiar song on the radio and think about the good times with dad and the old Corolla.

by Megan Ivancic [my loving daughter]

After such a touching and heartwarming tribute by my wonderful daughter, it got me to thinking about the power of sharing moments and a handout I would give to students attributed to Nadine Stair in 1953 when she was 85 years old.

If I had my life to live over again

I'd dare to make more mistakes next time.

I'd relax, I'd limber up.

I'd be sillier than I have been this trip.

I would take fewer things seriously.

I'd be crazier. I'd take more chances.

I would take more trips, I'd climb more mountains

I'd swim more rivers, I'd watch more sunsets,

I'd go to more places I have never been to.

I would, perhaps, have more actual troubles but fewer imaginary ones.

You see, I'm one of those people who was sensible and sane,

Hour after hour,

Day after day.

Oh, I have had my moments.

If I had to do it over again,

I would have more of them.

In fact, I would try to have nothing else but beautiful moments – moment by moment by moment.

If I had to do it over again,

I would have more of them.

In fact, I would try to have nothing else but beautiful moments-moment by moment by moment.

If I had to do it all over again,

I'd start barefoot earlier in the spring and stay that way later in the fall.

I'd ride more merry-go-rounds, I'd watch more sunrises. And I would play more with children, if I had to live my life over again.

But you see I don't.

By Nadine Stair in 1953

To me, that passage says it all. Life should be the creation of positive moments shared with family, friends, strangers who are friends you haven't made yet, co-workers and do I dare say students. With the entry my daughter wrote, I cannot help but feel totally blessed. If it is true that an acorn doesn't fall far from a tree, then I am so honoured even if she can be a bit of a nut from time to time.

THE IMPORTANCE OF A NATURE FIX

I am a big believer that nature provides soothing and healing for whatever ails you. It demonstrates wonder, beauty and can provide clarity to the things that truly matter in life. I would share a story how in my first year of teaching, I was feeling stressed and overwhelmed. My wife and I went cross country skiing in the backcountry by Maligne Lake in the Canadian Rockies.

The conditions turned a little for the worse when I noticed this massive wall of rock in front of me. I began to think of all the people who struggled to get through this pass over hundreds of years with inferior equipment and worse conditions and yet the rock was still there standing tall and looking majestic. It made me feel like my problems were quite trivial instantly putting me in a better head space and giving me the strength to put things in a healthier perspective. Here are some quotes to illuminate my point.

> *"Those who contemplate the beauty of the earth find reserves of strength that will endure as long as life lasts."* ~ Rachel Carson

> *"In all things of nature, there is something of the marvelous."* ~ Aristotle

> *"Plant seeds of happiness, hope, success and love; it will all come back to you in abundance. This is the law of nature."* ~ Steve Mariboldi

Never were these quotes better illustrated than during our grade six, week long camp excursions as so aptly and wonderfully described by my good friend Barry Gibson.

Grade 6 Camp

If you were in Grade 6, Spring meant one thing: Grade 6 camp. Surprise Lake was an element of the school, tracing back to the days the doors opened. Somewhere around the long weekend in May, all the Grade 6s, and some teachers piled into school buses and made the two hour trip west.

This particular camp offered a variety of outdoor pursuits from canoeing to forests to pond study. In truth, it mattered not what lessons the students had daily, the camp was more about how do you survive the better part of a week in the bush all the while learning to practice "being the best human you can be." It was not so much what content did you learn, but how you engaged with nature, and the rest of the folks on the journey with you.

In ways, the dining hall became the assembly hall where the teaching began and ended. Values were foundational to the experience: respect, kindness, cooperation, teamwork, friendship, loyalty, and competency. I often went as a Grade 6 home room teacher. My colleague, Art Muz, offered his wisdom from the back of the cookhouse, "We've been talking to these kids all year. Time for someone else to take the lead." And lead they did. In particular, Terry Halak and Mark Ivancic set a tone which we all then shared.

The truth of the matter was that many of these 11 and 12 year olds came with the excitement of going to a giant sleepover. For some, "let's get crazy" was a rough spot which needed some buffing. After all, you need your sleep, and six to eight people sleeping in a space about the size of a bedroom needs some real cooperation. A dining hall with 70 people needs some level of respect. There, students and teachers sat down with an emphasis on breaking bread with new people each meal. After lunch or dinner, we might share our experiences or sing some songs, building bonds. Each year, I watched, a rowdy **Monday** *crowd become a team working together by* **Tuesday** *evening. If it took until* **Wednesday,** *we'd view it as a tough group. The transformation from Me to We inspiring.*

Chores

Each day, you got a new set of chores. These varied from setting the table to washing the dishes to peeling vegetables to cleaning the biffies. Each task came with demands which taught both personal, camp, and societal goals. Most kids have never peeled a carrot, or learned the proper left and right of cutlery. Then there was the outhouse cleaners who came with their own mythology. Who will get suited up in a climbing harness to get lowered in for the thorough cleaning? Always someone like Peter was anxious to volunteer.

Mythology

Like any gathering of humans, tales must be told. From sidewinder snakes to a coyote skull to Arnold Poopchuck, stories were told to entertain and enhance the time together.

Transformational

Many kids came to camp with outdoor skills. They were comfortable with the physical, social, and emotional challenges. They often led, and in fact surpassed their achievements in a classroom. For others, this was a new experience. I still remember walking back into the camp from a good day in the bush and asking Pharyne, "How are you doing? Are you enjoying this?"

Pharyne turned to me and with great introspection said, "I don't know if you know this about me, but I'm not really a bugs and dirt kind of girl. This is the best time ever!"

Wrapping it all up **Thursday** *evening*

Here we are now, a team, and we can now celebrate our individual and group achievement. We had a few contests: fire lighting, canoe races, clean cabins. The ultimate in wrap up was THE MUD WALK. This began with some mythology and rituals: hand gestures (three fingers down to three fingers up- me to we), chants ("We are Spartans"), and songs (The Beverly HillBillies song complete with a hoedown). Then in worst the clothes they owned and shoes they wouldn't mind losing, students set off into the bush to a good stretch of bog. Mud was everywhere. Many volunteered to be tossed into a more watery part of the bog by Mark. Mr. I throwing kids was just part of the drama and practiced rituals. Returning to camp, the only choice was to dive into the lake to try to remove most of the mud. "Most" meant that some still found its way home.

Thank you

We returned to the dining hall for a final hot chocolate and a rousing edition of Crazy Train. ("Oh, it's **Thursday** night and I'm feeling all right").

We ended with a moment of thanks. Students and teachers were encouraged to express their gratitude to each other and then, we all returned to cabins one last time. It was a great moment, reaching out to shake hands, looking each other square in the eye, and together offering up our gratitude.

"Tomorrow morning we head home. Maybe I can have a shower to get rid of the rest of the mud. Good to see Mom and Dad. Boy, I have some stories to tell."

by Barry Gibson

THE POWER OF COMMUNITY

It seems I refer quite often to this theme even though it is not just important in the healthy development of children as I address in the various forms of programming our school had provided, but important for all of us. There are seemingly a lot of lonely people in the world today. With the advent to technology in our homes, we have a vast array of opportunities for stimulation without having to leave the confines of our own homes leading to less direct human interaction which is critical to our emotional well being. In poorer countries, it seems like there is more of a value placed on this as you will see families spanning several generations involved in some sort of meaningful activity whether it be work or play. More often than not suicide rates are considerably lower in these nations. Here are some quotes to support my claim on the importance of community.

> "One can acquire everything in solitude except character." ~ Stendhal

> "There is immense power when a group of people with similar interests gets together to work toward the same goals." ~ Idowu Koyenikan

> "I know there is strength in the differences between us. I know there is comfort, where we overlap." ~ Ani Difranco

> "We have all known the long loneliness and we have learned that the only solution is love and that love comes with community." ~ Dorothy Day

> "Alone, we can do so little; together, we can do so much." ~ Helen Keller

> "I am of the opinion that my life belongs to the whole community and as long as I live, it is my privilege to do for it whatever I can. I want to be thoroughly used up when I die, for the harder I work the more I live." ~ George Bernard Shaw

Being involved in a community does not have to be complicated because it comes in different sizes, shapes and forms as I have illustrated with the various school activities. It can be getting involved in activities hosted by your town or city, volunteer services, being a part of a club, sports team, choir, or simply doing things with people who have the same interests. When I was in the hospital and rehabilitation facility, I would tell anyone who would listen, how the hockey community was so good to me based on the number of visitors I received and how this was so crucial in my emotional recovery.

I was extremely hesitant in returning to my local gym being too self conscious in appearing all messed up and disabled. People were genuinely pleased to see me and these interactions totally buoyed my spirits even if they kept their distance from me when I was doing cable exercises with my damaged left arm not wanting to get harmed as the cables would inevitably go flying out of my weakened grip. It is another example of a social environment where the possibility of a positive interaction can and probably will occur which provides much needed nourishment for the soul. Not only that but you get to do your body a favour by working out. A lot of people are wary about going to a gym because they feel people are going to pass judgement upon them which was preventing me from going, but in actuality everyone is on their own personal mission for self improvement. When I switched the focus towards taking care of my own needs, it enabled me to move forward and I started to feel better about my situation.

The bottom line is, we all are in a happier place when we feel like we belong and we are part of a community. We have a lot of struggling youth trying to find their place and how they fit in. Yes, school is a community, but sometimes we need smaller communities within a much larger community just like in a town or city because, yes, it takes a village to raise a child.

Inevitably, I keep going back to the same theme, MAKE THINGS HAPPEN! For yourself and others if possible. In my first book, the focus was on My Road to Recovery Tour mountain hike.

I had thirty people join me and the focus was totally on me and what a powerful experience it proved to be for me and hopefully the others. I am making it an annual trip as I describe it as a fantastic community day, shared with old friends and new friends; because to date, everyone who has participated has brought great positive energy.

Except Dan during one trip. We noticed on two separate occasions, a black bear off in the distance. So on our way down, Dan in the spirit of, you don't have to outrun a bear, just the people you are with, says, "I am not worried if a bear crosses our path because Mark can't run!"

It is on the descent so I am feeling a little beat up and cranky. I am getting close to my old hockey mentality when I believed I was a tough guy thinking a new bear skin rug would look good in front of the fireplace. Attitude is everything so which bear would want to mess with someone who possesses that kind of mindset? Because even animals can sense the aura we exude. Obviously it worked because the bears stayed clear from our path! But let's be honest here, the more likely scenario would have involved a bear enjoying its new Mark skin rug as it enters its cave.

In essence, the trip has morphed into a splendid outing shared with wonderful people who enjoy a fantastic day together. And because there are repeat participants; presto, just like that I have created my own little hiking community filled with tremendous quality people. What a great day I get to relive each summer! And you are more than welcome to join in the fun providing you are willing to bring your best mojo.

BE A PEOPLE PERSON

This page was spurred by a conversation my wife and I had on our way to the doctor to get our results from the blood tests we had done. Now this can bring about anxiety for some people, but I remarked how I enjoyed interacting with our physician, an elderly gentleman at the twilight of his career.

After my stroke, I was skittish about seeing medical personnel, not wanting to hear any more bad news. This doctor comes across as if every one of his patients are so important and that you are in good hands. After every visit, I walk out of his office with buoyed spirits because he recognizes and reinforces how hard I have been working in my recovery.

He just oozes calm and kindness. When I mentioned how much I enjoyed him, my wife replied, "It's because he is a people person." As we enter the room, the first thing he says to us is, "It's the Bamma Ramma kids" referencing my first book I gave him.

Now right off the bat, by acknowledging it, I am humbled and pleased and he has set a positive mood for the rest of our visit. Another classic example of the effects of one small gesture and its impact.

So what is a people person? To me it signifies interacting with someone and whether it be for a minute or an extended period of time, you walk away in a better frame of mind thinking to yourself, I really enjoyed that. Surrounding yourself with people like that is crucial to your emotional well being. As you can tell by the people who provided submissions for this book, I am blessed to have quite a few in my life and if not, I fear I would be stuck in a dark place.

> *"When you are joyful, when you say yes to life and have fun and project positivity all around you, you become a sun in the center of every constellation, and people want to be near you."*
> *~ Shannon L. Alder*

Apparently, this is not everyone's cup of tea. I was running a parenting workshop and there was an older couple participating when the woman says, "Don't those people who think the sun shines out of their butt make you nauseous?" To which I replied, "Wait until you spend some time with me!" Of course, this incited me to amp up my performance which led them to leave and to never return. Another classic case of a win-win scenario.

MAKE SURE YOUR WORDS MATCH WHAT YOUR FACE IS SAYING

Have you ever had the experience where you meet someone and you think to yourself, "Wow, this person is really physically attractive." Then they start talking spewing negativity. After a while, upon further reflection when you look at them, you can't help but think, "Man, what was I thinking? This person really isn't really that attractive after all."

Ask anyone who knows anything about anything and they will tell you. "TRUE BEAUTY IS A WARM SMILE AND KIND EYES!" Hope for us all and it is a lot cheaper than plastic surgery.

Another teachable moment. In the previous quotation, you can even switch the adjectives around. Even though I am retired, it's moments like this I realize I still have some serious game left in me.

Kind eyes? I cannot help but think back to the time I was coaching and I received a technical foul without having uttered a word. When I tried to plead my case to the referee how I had not said a word. He replied how he did not like the way I was looking at him. He misconstrued my passionate basketball eyes as somehow I was passing judgement on him on what he probably believed was a fair and judicious performance.

I obviously held a conflicting opinion and another example in how differing opinions can create a negative scenario. And then it dawned on me, it wasn't my eyes, but this guy was good and he could read my thoughts not that you would have needed heightened paranormal skills at this particular time and place.

I tried to do the same with him when I realized he was like the guy who hears last call at the bar and wants to end the evening with a bang and is seriously considering ordering a double. A Double Technical without having said a word??!! If this had occurred, I am sure I would have felt compelled to retort with what I would have considered to be a witty and profound response such as, "Good job and perfect! One for each eye!" I am sure with the expression on my face, even my wife would not have considered me to be strikingly handsome at the time or the official would have believed me if I tried to explain to him how basically I was a nice guy.

But on a side note, wouldn't it be interesting if people could read our thoughts? Would it change how we view others and make us redirect our focus in a more positive direction? I am not worried because after a few minutes of reading my mind, I am sure the reaction would be, Borriiing!"or "That guy gots a funny brains." As my mom would say.

NELSON MANDELA

I find it mind boggling how we are in the year 2020 and racism is still an issue. Nelson Mandela ended apartheid; white dominance in his home country of South Africa. He demonstrated how it was possible to create positive change and establish equal rights for all in a non violent manner despite unjustly having to spend twentyseven of his, prime of life years in prison. It goes to show how sometimes in life we have to go through dark times in order to provide a light. If you are struggling, maybe it is what you need to go through in order to be the spark for someone else. Here is a man who we could all benefit from listening to and so I present these pearls of wisdom from a true hero and icon.

> "May your choices reflect your hopes, not your fears." ~ Nelson Mandela

> "It always seems impossible until it's done." ~ Nelson Mandela

> "Everyone can rise above their circumstances and achieve success." ~ Nelson Mandela

> "If they are dedicated and passionate about what they do. Just a reminder to be careful with the passionate eyes." ~ Nelson Mandela

> "There can be no greater gift than that of giving one's time and energy to helping others without expecting anything in return." ~ Nelson Mandela

> *"After climbing a great hill, one only finds that there are many more hills to climb."* ~ Nelson Mandela

I would just like to add, after climbing a great hill or overcoming an obstacle, you know you have it in you to tackle and be successful with the next one.

> *"Education is the most powerful tool which you can use to change the world."* ~ Author Unknown

What can I say? Once a teacher, always a teacher!

> *"What counts in life is not the mere fact that we have lived; it is what difference we have made to the lives of others that will determine the significance of the life we lead."* ~ Nelson Mandela

> *"The importance in self belief and a positive self image."* ~ Nelson Mandela

> *"What lies behind us and what lies before us are tiny matters compared to what lies within us."* ~ Ralph Waldo Emerson

I was speaking at a session with adults and I asked them to write on a piece of paper ten things they liked about themselves. I was surprised at how difficult a task this was for the majority of the room hearing things such as, "I need more time. I have six. Is that good enough? This is too hard." Yet when I asked them to write ten things they would like to change with themselves, it proved to be an easy ask where it seemed everyone finished in no time at all. Of course, when I asked fourteen year olds to do the same activity, the opposite occurred where the majority figured they were already cool enough with no reason to change.

As we age, it would seem that life's trials and tribulations take its toll on our self worth and how we view ourselves.

> "Owning our story and loving ourselves through that process is the bravest thing we will ever do." ~ Brene Brown

When we try to conceal our weaknesses, we draw more attention to them. I would use the analogy of being at a social event where everyone has to take their shoes off, and you notice your big toe sticking out of your sock. The instinct would be to try and conceal it, which would probably direct more attention towards it based on your awkward standing or sitting positions. It's like when men start doing comb overs when they start losing their hair and have bald spots. Yeah, like no one notices. Not that I would have any personal experience with this or do I?

> "Never bend your head. Always hold it high. Look the world straight in the face." ~ Helen Keller

> "You have been criticizing yourself for years, and it hasn't worked. Try approving of yourself and see what happens." ~ Louise L. Hay

I believe this next quote sums it up best.

> "The most beautiful people we have known are those who have known defeat, known suffering, known struggle, known loss and have found their way out of the depts. These persons have an appreciation, a sensitivity and an understanding of life that fills them with compassions, gentleness and a deep loving concern. Beautiful people do not just happen." ~ Elizabeth Kubler-Ross

Personally, this last quote really resonates with me as I struggled with my self worth after experiencing the debilitating results from my stroke; just as you may have your misplaced reasons, but we will get through this together.

> "The real difficulty is to overcome how you think about yourself." ~ Maya Angelou

> "You are always with yourself, so you might as well enjoy the company." ~ Diane Von Furstenberg

> "There is nothing noble about being superior to some other man. The true nobility is in being superior to your previous self." ~ Hindu Proverb

This next story is from my good friend and former colleague, Bob. I have given many examples of people overcoming obstacles in the world while leaving a lasting positive imprint and that is his goal in sharing his emotional journey. I am in awe of his courage and strength as you shall soon be. We discussed his participation on this project while hiking to the summit of Mount St. Piran at Lake Louise which demonstrates once again how being active with friends leads to all kinds of positive rewards. I knew he would do a good job, but his altruistic goal in desiring to help others dealing with anxiety is highly commendable as is his brutal honesty.

> "Although this may be a most difficult thing, if one will do it, it can be done. There is nothing that one should suppose cannot be done." ~ Tsunetomo Yomanoto Hagakure: 'The Book of the Samurai'

I retired from teaching at the age sixty. My reason for doing this was to help my incredible wife take care of my elderly parents. Carey was doing all she could, but the workload was becoming increasingly more difficult. My mother was already three years into her journey with Alzheimers and my father had just been diagnosed with dementia. As the millions of caretakers of loved ones with Alzheimers quickly realize, things never get better. There are not enough words to describe the cruel nature of this disease. Sally passed away two months after her 88th birthday.

> "Don't let me get me." ~ Pink

Two weeks after my mother's passing, I went to my family doctor to have a physical. We had a trip to Mexico booked for the next month and a physical was needed for travel insurance. Everything went well. A week later the doctor phoned and told me the electrocardiogram revealed I had Ventricular Ectopic Beats. I had ten extra beats per minute. He proceeded to tell me it was nothing to worry about, I was in good health. That night I became tense and irritable. After all, he did say I was ok so why was I so worried, stressed? You cannot control the events or circumstances of your life, but you can control your reactions. How often have we heard that in our lives? Toxic thinking was making its debut but and normal stress was about to go.

> "Sometimes it's lonely on the lost highway" ~ Van Morrison

A week after my doctor's call, my wife brought me to the Emergency; I was having my first panic attack. It was extreme and intense. My blood pressure was 178 over 120. I thought I was dying and about to have a heart attack. I was given a battery of tests which all came back positive. It was determined I might have White Coat Syndrome, where I, in effect, brought on the elevated blood pressure. Two weeks later I ended up in the Emergency again. Same tests, the doctors and nurses were doing their best. Everyone was so caring and professional.

The emergence of a panic disorder was taking place. I was having catastrophic misinterpretations of body or mental sensations. The releasing of cortisol (the stress hormone) caused an increase in heart rate and blood pressure. My fear and anxiety created a brutal cycle in which physical symptoms, emotions, as well as powerful negative thoughts, interconnected and escalated quickly.

I have had trouble sleeping all of my adult life. As you can well imagine, my attempts at sleep were marginalized therefore affecting my state of mind. After a couple of months I began to feel overwhelmed and hopeless. The feeling of heart palpitations seemed to be constant, catastrophic thoughts were often present, all symptoms messed with me on some level daily. I felt isolated and thought my condition was permanent. I was going it alone and I was spent.

> "A hand to take you to the light when it gets a little too dark to see" ~ Dave Loggins

My lack of sleep was playing havoc with me. I went four days with no sleep whatsoever. I had begun to have suicidal thoughts! One sleepless night I began to think of ways I might go through with this. On the fourth day without sleep, I returned home from helping two friends work on a motorcycle. Soon after settling in my chair I told my wife I was having suicidal thoughts and I wanted to die. She immediately phoned 911 and the ambulance came and they brought me to a hospital in the city.

From the very first moment in the hospital I was given hope. I spoke to a wonderful caring doctor who told me point blank that I could get better. I was admitted to the locked ward. Little did I know at the time, how transformational this would be for my life.

> "Patience is not passive, on the contrary, it is the concentrated strength." ~ Bruce Lee

At first, I spent time avoiding everyone. I stayed in my room as much as possible. Every morning I was scheduled to speak to the psychiatrist. It did feel good to speak to someone who knew exactly what I was going through. She was a great listener. As the nurses were taking my vital signs they always asked how I felt, they all showed such patience and caring. My sleeping was vastly improved and I was feeling more calm. I realized that my wellness was not on a trajectory. There would be good and not so good days. I was more patient than I had been in a long while, but something was missing to bring me to the next level.

> "When the student is ready, the teacher will appear." ~ Buddha Siddhartha Gautama Shakyamuni

One day I met a nurse who was assigned to me named Jeet. He began to tell me his observations about me, Jeet was very blunt and precisely on point. It was tough love, but I was buying it. I was over analyzing, he would say, "just breathe, don't believe everything you think." Jeet made me realize that I had to work on myself. It set in motion a newfound dedication to hard work, listening, and learning. What a genuine, knowledgeable person. This was not his first rodeo. I met with him for 15 mins every shift he worked.

The second person that would impact my life so positively was Kira. She was a nurse/therapist. Kira spoke to me about Cognitive Behavioural Therapy. She was passionate about me challenging my thoughts, moving me towards alternative balanced thinking. Kira was able to provide me with valuable information. She was a life saver. I read the material daily and worked very hard to learn and work the process. Kira appreciated my effort.

Finally there was Christine. She as well was a nurse/ therapist. Every morning I would work out in her small 'gym'. She was a big believer in the value of physical fitness. Her strength was dispensing timely info when she perceived it was needed. She was a great life coach.

> "When we learn how to become resilient, we learn how to embrace the beautifully broad spectrum of the human experience" ~ Author Unknown

One great attribute of resilient people is being able to ask for help. During a personal crisis, people can benefit from the help of qualified health professionals. They are experienced and prepared to help you, their skills far outmatch yours. Resilient people read about others who have experienced a similar situation. Read applicable books and journals. When you get the proper help, you begin to pivot in another direction, be ready to work and believe in yourself, no matter how bad it feels. My experience has taught me this.

By Bob

> "Ya Gotta Believe!" ~ Tug McGraw

I think this is the perfect time for the lyrics to the Beatles song HELP and try not singing it in your head.

*When I was younger so much younger than today
I never needed anybody's help in any way
But now these days are gone, I'm not so self assured
Now I find I've changed my mind and opened up the doors*

*Help me if you can, I'm feeling down
And I do appreciate you being round
Help me get my feet back on the ground
Won't you please, please help me?*

> *And now my life has changed in oh so many ways*
> *My independence seems to vanish in the haze*
> *But every now and then I feel so insecure*
> *I know that I just need you like I've never done before*
>
> *Help me if you can, I'm feeling down*
> *And I do appreciate you being round*
> *Help me get my feet back on the ground*
> *Won't you please, please help me?*
> *By the Beatles*

I find it most interesting how a song that was written in the sixties holds such relevance in today's world. The bottom line is, as Bob so eloquently described, it is okay to seek help when you are struggling and there are more options than ever before. I have seen a significant transformation with my friend and it looks really good on him but he needed to take the steps to help him get to where he needed to be.

BEING A STUDENT OF LIFE

The father of a girl I was dating as a young man once called me a student of life because I kept asking questions about other people's experiences. He kind of said it as an innocent observation without ever realizing how impactful those words would have on me throughout my life. This once again reiterates how the power of our words can have a huge impact on others both positively and negatively.

Everyone has a story to share and it is astonishing what you can learn from others that will help propel you further in life. Talking about themselves appears to be a favourite pastime of most people as it would seemingly appear based on my observations from Facebook. By asking questions, not only are you learning more about the other person and some valuable life lessons, but you are becoming more endearing in their eyes because you are displaying an aura that you care. When you think about it, without even knowing you, I would be willing to wager big bucks your best friends who you highly esteem are students of life. Unfortunately or probably luckily for my old flame, I was not a student in how to be a good boyfriend.

> "Life itself will protect you. You will graduate to living a better life than you have ever lived until now! Your patience and treatment has paid off! You are life's own student and have seen life from all angles. You have come full circle and now it is time to rejoice. Say you are absolutely fine and it will be granted to you!" ~ Sanchita Pandey, Cancer to Cure

> "Everything is teaching, if you are listening." ~ Anonymous

This next submission comes from my former student Justin who has morphed into being my good friend. He is one of those wonderful people who after you interact with you always walk away feeling a little more positive. The cool thing about this is he wrote his entry as a letter to his daughter and it really impacted and reinvigorated me when I needed it the most. It just goes to prove how positivity is the gift that keeps on giving.

> *A Letter to my Daughter,*
>
> *Last year in the fall of 2019, I embraced you on your way out the front door of our Calgary home. You were off to school to begin the 4th grade at your French immersion school, nervous about who your new teacher would be. I was on my way out the door in my 20th grade, halfway through my medical school training. I said to you that it wasn't so long ago when I entered my own fourth grade French immersion classroom and met my teacher Mark Ivancic for the first time, known more affectionately around Ecole Meridian Heights School as Mr. I. Now, dear daughter, I'd like to tell you more. I didn't know it then, but not only was I meeting the locally famous grade 4 French teacher, but also my future coach in sport and life, musical inspiration, bandmate, mentor, confidante, unceasing support, and my friend, who would lead me*

and so many others both literally and figuratively to the peaks of mountains.

Even at that young age, I was struck by how much Mr. I cared about each one of us individually in our bustling young class. He found ways to connect deeply to the one while simultaneously taking care of the many. I remember so fondly the first time he brought out his guitar and sang one of his songs for our class. The seed that would blossom for me into a lifelong love and passion for music performance was planted that day, particularly my love for playing guitar and singing. I know now it was my introduction to the magic of human connection through music, one I came to learn I couldn't live without. Mr. I exemplified care when discipline was necessary and embodied in every fibre of his being our school motto of "Be The Best You Can Be" while yearning for us to see that possibility in ourselves. All of this made my fourth grade a truly formative experience.

Over the next couple of years, Mr. I was always around ready to give a high five in the hallways or check in for a moment with sincere attention and encouragement. He filled school assemblies, gatherings, sports events and grade 5/6 camps with his hopeful presence, his uplifting music, and unfettered enthusiasm for life. He made me feel important, noticed, and anchored in a sea of developmental and social change. He likely never knew how important a role those small actions played in helping me when I was nearly swept away in the storm of my seventh grade. That year, at the age of 12 I was diagnosed with Type 1 Diabetes. My body no longer made insulin, a hormone critical to survival, and I was faced with the stark reality that I would be dependent on injections the rest of my life and must live with the looming threat of terrible complications if I deviated from the strict regimen required to maintain appropriate blood glucose control. I wondered if my hopes and dreams for a future were now dashed, and feared I'd be a pariah among my peers. Meanwhile, I struggled deeply with an inward war I waged between my orthodox Christian upbringing and finding my own identity entering adolescence. The encroaching sense of hopelessness gave rise to wishes that I would somehow cease to exist. I became very depressed, and I knew I was at a critical crossroads. I have never told him that the conversations we had during that time, so fortified with his pure belief in my ability to overcome and rise from that challenge, was a defining piece of my support network. I decided to stay. I was determined to pick myself up and become something so that I could one day help inspire others, like Mr. I had done for me.

That flame was further fueled by his grade 8 life skills class, designed to inspire the young generation to believe in themselves, to set our sights high in the sky, and through inspiring stories of trial and triumph instil within us important human virtues like kindness, humility, and service to others. I hung onto every word he spoke, truly believing that I could accomplish goals while being myself to the highest degree of authenticity. In my senior year of junior high Mr I invited me to try out for his senior basketball team. I had never played basketball in my life, due to focus on other sports. But he knew my heart and believed in my athleticism. I made the team. He became my coach, and he led us. We fought. We cried. We fell. We rose. We vanquished. There was no time for surface level with Coach Ivancic. He dove deep and pulled out the best in my teammates and I. Just like he had done in grade 4, 7 and 8, he did it again, and we won the championship that year. To top off my 10 years of formative schooling, Mr. I took the soon to be grads to scale a mountain peak together in an unforgettable capstone lesson on what we can accomplish when we work together.

These experiences propelled me forward into success in high school. I honed my craft with the guitar and voice and sought the company of good people who yearned to improve themselves and their world. Mr. I's support was ever-present, despite the less frequent connection. When I started a bike repair business in high school, he brought his bikes for tune-ups. He encouraged my decision to volunteer for two years on a mission and was there for *my farewell party. Prior to my departure, I brought my guitar to his home and sang and played a song for him and his wife, 10 years from the time he helped inspire the love of music in me. That was a catalyst moment to many joyful opportunities we would create in the future to collaborate in our mutual love of music. Upon my return, I joined Mark on stage for the first time to showcase his catalogue of thoughtful and moving songs from the past decade, in a band made up of former students. Sharing that stage was overwhelming, and as I took it in my memory flashed-back to myself as a quiet grade 4 kid seeing him play for the first time in class. He had made me feel like an equal then, and all these years later that hadn't changed, as our voices blended in a common passion for human connection through music.*

When I married at age 23, he didn't doubt I would rise to the challenge and unconditionally welcomed your mother with open arms. Over the years as you and your 3 siblings entered the world, Mark never missed an opportunity to vocalize how blessed my children were to have me as their father. In many moments of doubt, I clung to those words. He encouraged my academic pursuits, wherever they might lead me, and was one of the first to voice unequivocal belief and confidence in me after I applied to medical school, facing the daunting waiting period and uncertainty that threatened to overwhelm me. His utter lack of surprise when I received my acceptance letter spoke louder than any words. Around that time, I learned Mark suffered a devastating stroke. I was heartbroken in disbelief, but I knew it was my turn. I had to let him know how much I believed in him, that it was ok to be down. But not to stay down. That should sound familiar because my life was a testament to that. My life that he had been such an integral part of influencing. He didn't stay down. I don't know to what extent my interactions with him after the stroke were meaningful, but I see the evidence of the seeds he helped nurture in me as a young person bearing fruit in ways I never imagined. Perhaps some of that helped him cope, and may continue to do so as he sees that sweet fruit permeate into the way I raise my family, practice the *healing art of medicine, my music, my worldview, and the way I respect our planet. Humility, belief, humanity, respect, friendship, commitment, unconditional love and acceptance - these lessons ultimately transcended even the religion of my upbringing.*

So, my precious young daughter, as you come to know me, I want you to know that you also come to know my own fourth-grade French immersion teacher, my lifelong friend Mr. I.

Love Dad [Justin]

Trust me, after reading this, it lit a fire underneath me and had me more inspired than ever to stay the course and try to be a positive difference maker even though I sometimes question my ability to do so based on my current situation. A wise person once told me, we can all use a little support some time.

FINDING THE GOOD IN OTHERS

"Kindness is seeing the best in others when they cannot see it in themselves" ~ www.RandomActsOfKindness.org

"When you choose to see the good in others, you end up finding the good in yourself." ~ Sandelyn Kueh Lua

After reading Justin's take on his grade four experience, it takes me back to my favourite grade four lesson and yes, I stole it from a Chicken Soup for Soul book. I would talk about the power in kind words and how to deliver a heartfelt compliment. To demonstrate, I would try and give each student a personalized compliment. I am not going to lie because for the majority, the words just rolled off my tongue, but there were some words where I had to press and grind it out. But when you try and make this a regular practice, it is not that hard to do.

From here, I would get students to tape a piece of paper to their desk with their name clearly printed at the top of the page. Each student now had to work their way around the room and had to write one compliment for each of their classmates. They would always be so excited without even realizing they were practicing sentence writing. Contrary to popular belief, there was logic to my madness. When they returned to their seats to read what the others had written, their eyes sparkled with unbridled JOY!

I can't help but wonder what would happen if all organizations made this an annual practice. Would it impact how co-workers viewed each other? Would it improve productivity? With children, I noticed there was always a positive buzz in the room during this activity. The good usually lasted until the next recess, but there were always students who kept that paper taped to their desk for the rest of the year, re-reading it every day. What a great way to start a day!

Here we go again, homework time. During your next day at work, every time you come in contact with a different co-worker, see if you can formulate in your head something positive about them. If this assignment proves to be too daunting or difficult, we may have isolated where the problem lies.

It would seem whenever my daughter and I are having a conversation and the name of a former student comes up, my standard response is, "I really enjoyed him/her because he/she is a really good kid." To which my daughter will respond with, "You say that all the time, you are retired now, you don't have to feel obligated to make that statement."

Having spent years wiring my brain to see the good in all children, it is a genuine sentiment and one I needed to embrace to maintain my sanity during those difficult classroom moments. For me, it always made me feel better about myself and my career that I could make that statement.

Many people feel, finding the good in all others can be too difficult and taxing, but as with exercising, it becomes easier with regular practice and repetition. Trust me, I have been tested in this regard and there were many occasions where I wanted to wrinkle a student's neck [as my mother would say]. In those moments, I believed the student was making a bad decision because they did not know better and this was a teachable moment rather than passing some type of judgement on them. Sure, it is a lot more challenging when it comes to dealing with adults, but in believing the other person has good qualities and maybe they are not in a good place at the moment can help alleviate the drama in uncomfortable situations. And then let them read your book, this one that is, so they too can find a better path.

FEELING INFERIOR TO OTHERS

> *"No one can make you feel inferior without your consent."* ~ *Eleanor Roosevelt*

> *"Nobody is superior, nobody is inferior, but nobody is equal either. People are simply unique, incomparable. You are You. I am I."* ~ *Osho*

I was in Roger's office one day and he smiled and told me I was expletive recalcitrant. The first person I saw after this interaction was Doug, the head of our custodial staff. I shouted out to him, "Hey Doug guess what? Roger just called me recalcitrant," because I believed it had to be a good thing. Doug had this confused expression on his face like he had no idea what the word meant either or he was thinking to himself, "You aren't very bright, are you young man?"

I know, five years of university and I still don't know anything. One of the reasons why I wanted to include this story was that some people/teachers may have treated him like he was just the caretaker of the building when he was far more important than that. Each and every day he could be seen laughing and telling the same bad jokes with students. This made him an invaluable part of the school fabric; far more than some of the staff who felt they were superior because they had earned a piece of paper indicating they were intellectually advanced. Ask anyone who knows anything about the inner workings in a school and they will tell you it is the secretaries who keep the motor running. Like Coba, [our school secretary] who always had a bag of candies in her desk. Every morning you would see a crowd of waifs all around her receiving a treat and a much needed hug to start their day. Just a secretary? Right!

In our school, we were blessed with so many wonderful and caring individuals who comprised our custodial staff and secretary pool. Each one contributed in their own special way, making our school a warm and safe place for children. Their contribution was and is immeasurable.

It is like during parent/teacher interviews when a mother would say, "I am just a stay at home mom." Really? Do you need to include the word 'just' in that sentence? In other words, you are 'just' focusing all your attention towards arguably the most important role in our lifetimes, the raising of children.

The point I am trying to make is, regardless of whatever position one holds in life, it is insane to put the word 'just' in front of it and somehow lower your value or worth. We can all make a positive difference, which is the moral of the story.

For those of you who are curious and would like to know, here is the official definition of recalcitrant: having an obstinately uncooperative attitude towards authority. Sure I may have bent the odd rule occasionally, but only if I thought it would be of benefit for a student. I would like to believe I was respectfully recalcitrant. And here we go again. Respectfully recalcitrant... happy death... educated redneck.

What I did have a problem with was when people in positions of authority would start a sentence with, "As your superior, I would suggest, blah blah blah." If you have to start a phrase with those words, it demonstrates a different type of need. Every successful team I ever played on or coached; the best players were the best people, not expecting any special or preferential treatment. They valued the contributions of everyone and brought out the best from their teammates. They understand each person added their own unique talents towards the overall success. Which does not explain how, in my final year of university when I was named team captain, our team did not win a single game during the entire season. By no means was I the best player though.

It goes without saying, humility and respect for all would be the qualities exhibited by the leaders in most successful organizations. Feelings of superiority is the backbone of racism and a cancer within any group. When you look at the friction between different religions, it usually starts when one group has what they consider to be a superior understanding with their interpretation of the teachings in the bible, the ultimate book on hope, love, resilience, compassion and understanding.

I am just going to end this recalcitrant rant with YOU ARE NOT ANY LESS OR MORE IMPORTANT THAN ANY OTHER INDIVIDUAL ON THE PLANET!

LIFE IS A GAME

Life is a game ? Sure it is, just go to the board game aisle in any toy store and you will see it on a shelf. Except in the real game of life, the outcome is not determined by some random roll of the dice and there does not have to be winners and losers. An old sports adage. *"SOME PEOPLE LIKE TO WIN, SOME LOVE TO WIN AND SOME NEED TO WIN."* And I would offer, we all need to and can win despite our differences.

In coaching basketball, my most successful teams always had players with differing strengths, but when combined together in unity, we enjoyed a lot of victories and collective joy.

> *"Life is a puzzle, a riddle, a test, a mystery, a game-whatever challenge you wish to compare it to. Just remember, you are not the only participant; no one person holds all the answers, the pieces, or the cards. The trick to success in this life is to accumulate teammates and not opponents."* ~ Richelle E. Goodrich.

Unfortunately, despite how well you play the game, you are going to rub some people the wrong way, but that should be overshadowed by the number of great teammates, friends and quality people you have in your life. We all have them and I consider it a true blessing for all the people I have met and enjoyed moments with in my life. Bravo for you if you feel the same way.

> *"I have come to realize that life is neither a battle nor a game to be won, it is a game nonetheless, but to be played... enjoyed. There are neither winners nor losers... just players and what's great is that you can choose who to play with."* ~ Val Uchendu.

The importance of good teammates, my oh my, I just found another topic to give an opinion on. The key to playing any game is to have fun and the more skilled you get at it, the more enjoyable it is. I do not play computer or video games, but who enjoys being stuck on level one? Another old sports adage:

> *"You need to learn how to lose and respond to losses before you learn how to win."* ~ Author Unknown

No one said it was going to be easy and here, I am going to let you in on a golden secret formula passed down from the ancient sages, it takes HOPE, LOVE AND RESILIENCE.

It is 2020 and with all the seeming craziness in the world, hopefully it is just a wake up call to remind us that we are all in this game together and each and everyone of us need to step up and be better teammates.

EDUCATION

> *"Education is the most powerful weapon which you can use to change the world."* ~ Nelson Mandela

> "Learning is not compulsory. Neither is survival." ~ Dr. W. Edwards Deming

> "Don't forget to read inspirational math quotes when things just aren't adding up." ~ Anonymous

> "Education gives us an understanding of the world around us and offers us an opportunity to use that knowledge wisely." ~ Anonymous

> "Education is the passport to the future, for tomorrow belongs to those who prepare for it today." ~ Malcolm X

> "Education is what remains after one has forgotten what one has learned in school." ~ Albert Einstein

> "The more that you read, the more things you will know, the more that you learn, the more places you will go." ~ Dr. Seuss

Having spent the majority of my life involved in education either as a student or a teacher, it is quite obvious to see how this area holds special significance for me , but we are all in class each and every day.

> "The learning process continues until the day you die." ~ Kirk Douglas

> "Live as if you were to die tomorrow. Learn as if you were to live forever." ~ Mahatma Gandhi

For someone who has been downplaying the relevance of opinions, it would seem I have been sharing more than a few of my own. Even if I tried, I do not have the strength nor desire to hold back in this department.

When I think of teaching, I always harken back to a story Roger would share with the staff which happened during his first year of teaching. Apparently his principal thought he was not taking the job as serious as he should have been and said to him, "ROGER, YOU GET YOUR HEAD OUT OF THE CLOUDS AND DON'T YOU EVER FORGET THAT TEACHING IS THE MOST NOBLE PROFESSION OUT THERE!"

Even though it feels like it has lost its luster and sheen in the eyes of many. The greatest resource in any culture is its youth and what a noble venture and expedition it is to have the opportunity to help mold and empower them. The majority of teachers I know share a common trait, they are giving people. And some of the recalcitrant readers are now probably mumbling to themselves, "Yeah right."

The givers of detentions, the givers of too much homework and the givers of boring classes. "I could go on, but let's stay focused with the theme of this book. Teaching children can be comparable to riding an emotional roller coaster as many parents found out during the pandemic, but it does bring incredible rewards.

Teaching can prove to be most challenging in today's world as many educators embrace the notion that they have to be the solution. As I addressed earlier, the best anyone can hope for is to be a positive part of the solution as it is a team undertaking. There is an underlying expectation a teacher must meet the needs of every individual they come in contact with and they are the master educator in every subject they teach; keeping up with all the latest trends.

Which is understandable if you are a specialist, but if you are instructing multiple subjects, it can be a bit much. Once again, like in every successful organization or team, it takes a collective group based on individual strengths. I had some definite areas of weakness, but I always figured there would be someone else down the line who could provide the quality programming that maybe I did not and vice versa with my strengths.

I have to say, as positive as their submissions were towards me, I am shocked to hear how two remarkable students/students, Justin and Mary went through such struggles. They were and continue to be movers and shakers which makes me wonder how many others are quietly going through extremely difficult personal struggles without anyone knowing?

I was chatting with a colleague who went to a different school and he mentioned the high volume of students he felt were seemingly battling with depression and he thought one of the primary reasons was a social disconnect. Which is why we need more outside of the box type of thinking regarding meeting the needs of students. It would be ludicrous to put any more on a teacher's plate.

What I am proposing, is that school divisions invest as much time, effort and funding as they do towards technology, towards the creation and implementation of programming that match teachers' strengths and passions with those of students. What is the value in producing high academic results if there is apathy and the kids don't care?

There are those who embrace the ideology that a school's role is solely to focus on educational development rather than an opportunity and responsibility to develop well-rounded people. They treat the curriculum as if it was etched in stone and that Moses carted it down Mount Sinai in a wheelbarrow. They are important guides and I value the importance of a good guidebook after having gone on the wrong route on many mountain scrambles which often led to the most memorable and perilous moments.

I was going to quote Frank Sinatra, but in this case it would be, "too many to mention." I see no value in the mastery of a curriculum if a student cries themselves to sleep each night. I have had some wonderful young people confess this to me. So allow and accommodate the development of a diverse range of programming so students with the same interests can gravitate to it where they can learn and develop skills to help them grow into a more rounded individual.

This takes me to my great friend and former hockey teammate, Kayee, who is a wizard in building things and developed a construction program for high school students. They would land contracts and go out in the community and build things. Not only were students learning hands-on skills, but other lessons as well, like taking pride in a job well done and developing a hearty work ethic. Very powerful! I have offered many examples of the programming we were involved with which no longer exist today, due to a variety of valid reasons (or so we think).

For example, someone decided my Life Skills classes were no longer necessary which shut down a lot of the other programs I offered. Initially, I was quite bitter with this decision, but by trying to adopt an attitude of gratitude I now consider it a blessing because it left me feeling like I had unfinished business. Which was one of the key factors in the creation of this book and it has proven to be pivotal in my emotional recovery from my stroke.

This quote from one of the greatest minds in the history of mankind, Aristotle, sums it up best, *"EDUCATING THE MIND WITHOUT EDUCATING THE HEART IS NO EDUCATION AT ALL."* And there you go, I have now officially demonstrated how I share the same thinking with one of the greatest philosophers of all time. Once again, it harkens to being part of a community and the formation of new teammates.

> *"Education without values, as useful as it is, seems rather to make man a more clever devil."* ~ C. S. Lewis

One of my personal dreams with this book is that there would be teachers who share it with their students and one of the them becomes an educator down the road and shares it with their class. Now that would be a definite bamma ramma slamma jamma moment!

The majority of the teachers I know are caring individuals who are trying to do what they believe is best for children and they are definitely not infallible. We all have to remember we are all on the same team trying to do what is best for the child. Children are resilient and this needs to be reinforced, instead of running to their aid to bail them out when they have erred.

Political humourist Bill Maher summed it up best when he said, *"THE DECLINE OF THE AMERICAN EMPIRE STARTED WHEN PARENTS STARTED SIDING WITH THEIR KIDS OVER THEIR TEACHERS."* We have all had teachers who were seemingly unkind to us and as unpleasant as it was, truth be told, we all have had the pleasure of having teachers who saw the potential in us and believed in us. Who do you want to focus on?

When there is an issue, we can use these experiences to teach life lessons to our children demonstrating the power of healthy communication and dialogue. We are all going to learn how to deal with conflict with authority figures and especially if we are respectfully recalcitrant. Maybe we need to get back to a time when we got in trouble at school, it was worse when we got home, which takes me back to when I was about six years old.

Whenever we were being consequence as children, my mother would make us kneel in a corner in the living room. Apparently I came home from school one day after having made a bad decision and went directly to the spot which was clearly delineated by now because of the misdeeds of my siblings. My mother walks into the room and is surprised to see me in this position. I, believing there was a special telepathy between parents and teachers, was just trying to avoid the rush as my brother and sister would be soon coming home.

Oh yes, the good old days. Looking back. I deserved more time in that corner. Children are resilient and we need to build and foster this quality so they can have it when they really need it as they get older. When we regularly throw life lines to protect our children, whose' needs are we meeting?

I know, Tears of Jupiter, your best friend standing up for you even when I know you are wrong. We can use this as an opportunity to teach conflict resolution skills instead of a rallying cry to get the pitchforks and torches out to destroy the evil menace. It is only my opinion anyways, but the education of our children necessitates a team approach in today's world. Let's face it, we are all on the same side or at least we should be, rather than playing the blame game which can only result in divisiveness. Remember, we don't prepare the journey for the child, we prepare the child for the journey.

This seems like the appropriate follow up.

TEAMWORK

> "Teamwork: Simply stated, it is less me and more we." ~ Anonymous

I was watching an interview with hall of fame baseball player and manager, Dusty Baker. He referenced a conversation he enjoyed with basketball legend Bill Russell. A man whose Boston Celtics teams won an amazing eleven championships in thirteen years. Dusty asked him, what was the key to all the success? Was it coaching? Was it a system? Certain players? And Bill Russell replied, "No man. We loved each other!" And this was in the late fifties and early sixties when racism was supposed to becoming a thing of the past. And this team was a shining example as it was composed of whites and blacks who demonstrated great harmony and chemistry on and off the court.

I had always coached boys basketball which changed when I taught a specific grade four class. The girls were hard working, motivated, good athletes, loved to read, were the kindest and most empathetic people in the room. They were basically really good people; so I knew I had to coach them in junior high.

I am frequently reminded by them, the speech I gave when I announced the final selections, "Girls, meet your new best friends." Time surely does fly, but it is now close to twenty years later and many are still best friends. And did we ever win a lot of basketball games!

We were in a small community on the outskirts of the city of Edmonton and my previous teams had little success against the city teams. They were being intimidated with the belief we were just yokels from the country, playing against the sophistication of urban street ballers. But that all changed with this team who played their hearts out for one another.

For three years we went into unchartered waters and became a force and it changed our basketball culture for years to come. We gained a bit of a reputation and the players coming up believed it could be done as we enjoyed a fair bit of success for years after. It all started because I had a team loaded with compassion, character and love for one another. This can be applied to any community or organization we are connected with. It is easily within us to perform daily small positive gestures to enhance the overall culture and to contribute as part of the solution making it a more successful and enjoyable environment. The only problem with this team was that their basketball skills evolved faster than my coaching skills.

> "When he took time to help the man up the mountain, Lo, he scaled it himself." ~ Tibetan proverb

> "None of us, including me, ever do great things. But we can all do small things, with great love, and together we can do something wonderful." ~ Mother Teresa

EVOLVE OR DISSOLVE

> "Life is a series of natural and spontaneous changes. Don't resist them; that only creates sorrow. Let reality be reality." ~ Author Unknown

I heard from my friend Pete who once told me, eighty percent of reality is perception and eighty percent of perception is reality which means we get to choose how we see the world. I would like to believe this is true.

> "Let things flow naturally forward in whatever way they like." ~ Acclaimed Chinese philosopher, Lao Tzu

> "Loving people live in a loving world. Hostile people live in a hostile world. Same world." ~ Wayne Dyer, Best Selling Author, Motivational Speaker

Well, I had a spontaneous change in my life and I struggled to get back into the flow. I was not a happy camper when all the activities which always brought me great joy were suddenly and inexplicably taken away from me. Some people who knew how much I enjoyed an active lifestyle would echo my sentiments by saying, "It makes no sense, YOU of all people!" While others on the expedition team would try to be encouraging by saying, "I guess you will just have to find new things to be passionate about."

I would mumble in response, "If there were other things I wanted to be passionate about, I would have tried them before I had a stroke." I cannot believe my next words are in print because it is extremely improbable you would ever hear me say them in person. In hindsight, "You were right and I was wrong." And that is what I did, I started looking for new ways because I had to evolve and change to make me feel like I was back in the game.

I always sensed I had a book in me to write, but now I discovered a new found purpose. It has proven to be a most challenging and stimulating process with unforeseen benefits. Just like all expeditions.

> *"I can be changed by what happens to me. But I refuse to be reduced by it." ~ Maya Angelou*

I have a new three wheel recumbent bike and can be seen cruising around town on almost a daily basis. Hopefully, conditions will be settled by next summer because I plan to bike and camp from Jasper to Banff on one of the best scenic and bamma ramma slamma jamma roads on the planet. I'll head through the Canadian Rockies with an extraordinary team already in place, Souchie, Dan, Kelly, Matt and Barry, but this is subject to change if they do not come to terms with the fact that I am their superior on this expedition!

I mean someone has to step us and lead the charges. Otherwise, it will be just like herding cats. This trip is one of the main catalysts for the title of the book as it surely is going to be a prime example of a Hope, Love and Resilience Expedition! Hope as in, I hope we get the chance to do this and if not this, then definitely something else. Or that I hope I am physically capable of doing this and for those who question my resilience, we can no longer be friends even if we haven't even met yet. Or as in love, like I am going to love the incredible views along the way, all the laughter we are going to enjoy and the feeling of accomplishment when all is said and done.

Now I do scrambles I normally would not have because I had previously categorized them as not challenging enough. Well they definitely are challenging now in a much different way! And that brings joy to me. I guess after all this, what I am trying to say is, what we all need to do is find our way back to the flow when unpleasant change happens. No truer words have been written in this book:

> *"When you are hit with life-disrupting events, you will never be the same again. You either cope or crumble, you become better or bitter; you emerge stronger or weaker." ~ Al Siebert*

> *"Those who expect moments of change to be comfortable and free of conflict have not learned their history."* ~ Joan Wallach Scott

Is history not another school subject? So pay attention kids!

> *"To exist is to change, to change is to mature, to mature is to go on creating oneself endlessly."* ~ Henri Bergson

I had a hard time writing this quote, imagining all the snickering from former students. "You? Mature? That's funny. You always were a silly man."

> *"All changes, even the most longed for, have their melancholy, for what we leave behind us is a part of ourselves; we must die to one life before we can enter into another."* ~ Anatole France

> *"To be nobody but yourself in a world that's doing its best to make you somebody else is to fight the hardest battle you are ever going to fight. Never stop fighting."* ~ E.E. Cummings

Well I stopped fighting on the ice and trust me that was a good thing.

> *"I cannot say things will get better if we change; what I can say is they must change if they are to get better."* ~ George C. Lichtenberg

I once read where the human psyche is designed to handle change at a slow and somewhat controlled pace and that is not happening in today's world where change is seemingly a regular occurrence. Which is why it is critical in today's world to live in the moment and not let the torrent of change make our river swell to the point we get out of the desired flow. When change does blindside us, that is when the focus on HOPE needs to kick in. Not just the hope things WILL get better, but the hope and belief in oneself that you can and will find a way to improve the situation because if not you, then who? I am not going to lie, I got lost for a while which is an easy enough thing to do and it did me little good.

> *"I think it's a mistake to ever look for hope outside of one's self."* ~ Arthur Miller

> *"I am down and that is okay. I may be down for a while, but I will rise again. And when I rise, I will rise higher than I have gone before, I will be stronger than I've been before. I will thrive."* ~ Brian Vaszily, Author of 'The 9 Intense Experiences'

> *"What is hope but a feeling of optimism, a thought that says things will improve, it won't always be bleak, there's a way to rise above the present circumstances. Hope is an internal awareness that you do not have to suffer forever, and that somehow, somewhere there is a remedy for despair that you will come upon if you can only maintain this expectancy in your heart."* ~ Wayne Dyer

> *"Hope is important because it can make the present moment less difficult to bear. If we believe that tomorrow will be better, we can bear a hardship today."* ~ Thich Nhat Hanh

> *"Don't brood. Get on with living and loving. You don't have forever."* ~ Leo Buscaglia

> *"Our human compassion binds us one to the other-not in pity or patronizingly, but as human beings who have learnt how to turn our common suffering into hope for the future."* ~ Nelson Mandela

> *"Optimism is the faith that leads to achievement. Nothing can be done without hope and confidence."* ~ Helen Keller

And every time we take a little action to improve ourselves or the situation, we plant the seeds for more hope and a new found inner strength."

> *"The journey of a thousand miles begins with one step."* ~ Lao Tzu

> *"They say a person needs just three things to be truly happy in this world; someone to love, something to do and something to hope for."* ~ Tom Bodett

Sounds like a promo for a hope, love and resilience expedition or an invite to the road recovery tour.

Speaking of the Hope, Love and Resilience Bike Expedition. I mentioned it to my great friend, Kelly, and his first words were, "I want in!" Now Kelly is the type of friend we all need. He is the kind of person you may not see for an extended period of time, but when you get together you just naturally slide back into the groove where you last left off and it is an enjoyable smooth transition. I had a separate conversation with his son Matt, who was thrilled to be part of the expedition. You know what they say, the acorn does not fall far from the tree, and yes they both are a bit nuts!

Kelly and I started talking about the logistics of the trip because there are a lot of variables in play which reminded me of what I would constantly tell students, "THERE AREN'T PROBLEMS

ONLY SOLUTIONS." I told him no matter what crops up we will work it out and find a way. On the topic of solutions, I offer these because sometimes we can overthink things rather than believing you are flexible and will be able to navigate your way through unpleasant circumstances that unexpectedly arise.

Simply put, sometimes things turn out exactly like we plan and then there are times we have to adapt to the situation and make the best of it knowing there is always a path to the desired end. Future brides should pay attention to this and I am wise enough to not infer that this has anything to do with my engaged daughter.

> *"There is no easy want that can resolve our problems. The solution rests with our work and discipline."* ~ Jose Eduardo dos Santos

> *"When I am working on a problem, I never think about beauty but when I have finished, if the solution is not beautiful, I know it is wrong."* ~ R. Buckminster Fuller

> *"Suicide is a permanent solution to a temporary problem."* ~ Phil Donahue

> *"There's a lot of beauty in the world, so go hang out and be part of the solution rather than the problem."* ~ Mac Miller

I guess what I am trying to say, is that life is like an EXPEDITION. Sometimes gruelling, but you have fellow teammates with the same common goal; quest for inner peace, happiness and a little adventure along the way, who help you get through the grind. And remember, just like my hope, love and resilience bike expedition, you are the superior figure head on this journey which means you get to decide who accompanies you and what the adventure is going to be.

My Dear Son,

I am so very proud of you. Now, as you embark on a new journey, I'd like to share this one piece of advice. Always remember that adversity is not a detour. It is part of the path.

You will encounter obstacles. You will make mistakes. Be grateful for both. Your obstacles and mistakes will be your greatest teachers. And the only way to not make mistakes in this life is to do nothing, which is the biggest mistake of all.

Your challenges, if you let them, will become your greatest allies. Mountains can crush or raise you, depending on which side of the mountain you choose to stand on. All history bears out that the great, those who have changed the world, have all suffered great challenges. And, more times than not it's precisely those challenges that, in God's time, lead to triumph.

Abhor victimhood. Denounce entitlement. Neither are gifts, rather cages to damn the soul. Everyone who has walked this earth is a victim of injustice. Everyone.

Most of all, do not be quick to denounce your sufferings. The difficult road you are called to walk may in fact be your only path to success.

Richard Paul Evans, *A Winter Dream*

"*Life is a journey between perception and reality!*" ~ Ramana Pemmaraju

---

That quote has to be true because the author has part of bamma ramma slamma jamma in his first name.

"*The best part of life is to decide to make the journey through life like a best selling book. Tell a fantastic story to tell others how you did it.*" ~ Catherine Pulsifer

Even though I am writing my second book, I firmly believe there are still many more chapters to add to the story of my journey and expedition which have yet to come and will only occur if I choose to make it happen.

> "My grandmother started walking five miles a day when she was sixty. She is ninety seven now, and we don't know where the hell she is." ~ Ellen DeGeneres

She was probably on a fitbit challenge. So be careful kids.

> "Look, if you're driving down the highway at 120 miles an hour, I'd rather be behind the wheel than in the backseat." ~ Mark Wahlberg.

Because you need to be the driving force on your expedition!

> "You must travel you own journey at your own pace, do not rush the process... or assume you have to accept someone else's choices for your life." ~ Thomas L. Jackson.

I really like this because once again it reiterates how it is your journey and how you know you best and maybe you will get clarity when you are walking down the hill.

> "Take your journey all the way. Don't just do what you want to do. Do not pick and choose what you have the strength to deal with. Take all of it and to everything possible to change your life." ~ Therese Benedict

Things happen to us which we wish never did because I would never wish for a stroke to happen to anyone, especially someone I know. But here I find myself, post stroke, writing a book and opening the door to new possibilities and experiences. And none of this could have been possible without the creation of an optimistic outlook. There were definitely dark times where I figured things would have been so much better if I had just died.

> "Optimism is the faith that leads to achievement. Nothing can be done without hope and confidence." ~ Helen Keller

> "It's not that optimism solves all of life's problems; it is just that it can sometimes make the difference between coping and collapsing." ~ Lucy Mac Donald, 'Learn to Be an Optimist'

And for my peers:

> "You are never too old to set another goal or to dream a new dream." ~ C. S. Lewis

Once again, this strikes a chord with me. Pre-stroke, I felt like a vibrant and dynamic personality full of dreams and potential who aged fifty years overnight! Wow! That would make me really old! I felt like I became a broken old man who lost his mojo.

I now have new goals and dreams which makes my thinking of the future much more exciting and positive with unique potential. For all of us, tomorrow can and will bring more laughter, heartfelt moments and unexpected joy. It is up to us to make it happen and to embrace whatever obstacle that crosses our path on the expedition.

> "The optimist turns the impossible into the possible, the pessimist turns the possible into the impossible." ~ William Arthur Ward

Miracles are events that are seemingly impossible that happen anyways.

> "Optimists are right. So are pessimists. It's up to you to choose which you will be." ~ Harvey Mackay

In my case, having gone through both, it seems being optimistic is a lot more fun!

> "The difference between an optimist and a pessimist? An optimist laughs to forget, but a pessimist forgets to laugh." ~ Tom Bodett

This is my rationale for my trying to be humorous throughout the course of this book. In my mind, if you did not find my attempts amusing, it is probably on you and not me! Or maybe I am just not that funny. Once again a misuse of the word 'just' because sometimes we simply have to entertain ourselves.

To think, I wanted to have 365 entries in this book! I think it is time to wind it down in style with the wisdom of Frank Sinatra...

> ***MY WAY***
> *And now, the end is near*
> *And so I face the final curtain*
> *My friends, I'll say it clear*
> *I'll state my case of which I'm certain*
> *I've lived a life that's full*
> *I travelled each and every highway*
> *But more, much more than this*
>
> *I DID IT MY WAY*
>
> *Regrets, I've had a few*
> *But then again, too few to mention*
> *I did what I had to do*
> *And saw it through, without exemption*

I planned each chartered course
Each careful step along the byway
But more, much more than this

I DID IT MY WAY

Yes, there were times, I'm sure you knew
When I bit off more than I could chew
But through it all, when there was doubt
I ate it up and spit it out
I faced it all and I stood tall

AND DID IT MY WAY

I've loved, laughed and cried
I've had my fill, my share of losing
And now, as tears subside
I find it all so amusing
To think I did all that
And may I say, not in a shy way
Oh no, no, not me

I DID IT MY WAY

For what is a man, what has he got
If not himself then he has not
To say all the things he truly feels
And not the words of one who kneels
The record shows, I took the blows

BUT I DID IT MY WAY

By Frank Sinatra

I cannot help but feel these should be the words we all sing as our final swan song. I was listening to Frank Sinatra singing this song when it dawned on me how this should be an ebook with a musical soundtrack. With that in mind, many of the songs I referenced have tremendous lyrics with powerful messages. So as the curtain closes on this book, I would like to wrap things up with a few final observations.

On the Joe Biden campaign trail, Barack Obama is filmed playing in a pick up game of basketball where he nails a step back three pointer and yells to the camera, "THAT'S HOW I DO!"

Now some may interpret that statement as a tad cocky, but to me, it represented someone celebrating in the moment and enjoying himself. He had reason to be excited after hitting nothing, but not with the cameras rolling! It should be okay to celebrate big moments in life.

For those who feel I may have been a tad self-serving with this book, I would respond with, "That's how I do and how I did!"

By no means would I consider myself a master teacher who positively impacted every student I interacted with, but I did try to be part of the solution which is the best we all can hope to be. I did enjoy a remarkable career filled with much laughter and bamma ramma slamma jamma moments with tremendous young people from wonderful families. Even though I feel like I retired before my best before expiry date and not the way I envisioned. Looking back, it was all on me.

The next time this situation plays out, I will do better because I now know better! Next time? I will get back to you, if someone chooses to hire me for the next thirty years, to let you know if I handled things better, but who am I trying to kid? I will still be respectfully recalcitrant and try to do things my way. So let this book be the final lesson I give in my educational career. Now that's how you walk away from the game, with a mic drop and a little Connor McGregor swagger.

I would just like to wish you all the best on your expedition. May it be filled with great music, shared with a BFF. I don't understand why people would not substitute 'a' with 'the'. I would assume, and you know what they say about assume; you realize I win the title for the best friends' award because I assume if you are reading this, you have read the entire book and you have remarked how there were some pretty powerful submissions from impressive people who are my friends. And yes that is better than awesome! with an abundance of LOVE, HOPE AND RESILIENCE AND AN ENDLESS SUPPLY OF BAMMA RAMMA SLAMMA JAMMA MOMENTS and do not forget to do it your way as only you can do!

ABOUT THE AUTHOR

Mark Ivancic's life is an adventure. Whether teaching, strumming his guitar, playing a game of hockey, or climbing a mountain, Mark faces life with energy and optimism. He has this desire to grow stronger mentally, physically, emotionally, and spiritually. His passion is to share the lessons of life with those around him.

Mark was born in Shumacher, Ontario to proud Slovenian immigrants, Joe and Josephine. His journey then took him to St. Thomas University to play hockey, on to Windsor to work with at risk youth, and then to Alberta to be closer to teach (and closer to the Rockies). He and his wife Ann Marie welcomed Megan to their home and parenting became a new passion. For 28 years, Mark taught French Immersion as well as taking a leadership role with junior high students offering them opportunities to grow their humanity through Life Skills courses.

Drastic life change came in the spring of 2018 when Mark returned home from a hockey game and suffered a stroke. Facing mental and physical challenges greater than any he had ever known, Mark began a journey that continues to this day. It is a journey of hope, love, and resilience.

Manufactured by Amazon.ca
Bolton, ON